# METEORITES

## The story of our solar system

### SECOND EDITION

**Caroline Smith, Sara Russell & Natasha Almeida**

FIREFLY BOOKS

# A Firefly Book

FRONT COVER: © The Esquel pallasite, composed of gem-quality olivine crystals embedded in metal. © Natural History Museum, London

BACK COVER: Aerial view of the Pingualuit Crater in Quebec, Canada, a meteorite impact crater. © Courtesy of Denis Sarrazin, NASA Earth Observatory

Published by Firefly Books Ltd. 2019

Copyright © 2018 The Trustees of the Natural History Museum, London,

First printing

**Publisher Cataloging-in-Publication Data (U.S.)**

Library of Congress Control Number: 2018956220

**Library and Archives Canada Cataloguing in Publication**

Smith, Caroline, 1976-, author
       Meteorites : the story of our solar system / Caroline Smith, Sara Russell & Natasha Almeida. -- Second edition.
Includes bibliographical references and index.
ISBN 978-0-228-10174-1 (softcover)
1. Meteorites.  I. Russell, Sara S. (Sara Samantha), 1966-, author II. Almeida, Natasha, author  III. Title.
QB755.S64 2019                 523.5'1                 C2018-905090-X

Published in the United States by
Firefly Books (U.S.) Inc.
P.O. Box 1338, Ellicott Station
Buffalo, New York 14205

Published in Canada by
Firefly Books Ltd.
50 Staples Avenue, Unit 1
Richmond Hill, Ontario L4B 0A7

First published by the Natural
History Museum, Cromwell Road,
London SW7 5BD

Designed by Mercer Design, London
Reproduction by Saxon Digital Services UK
Printed by Toppan Leefung Printing Limited

# Contents

# CHAPTER 1

# What are meteorites?

METEORITES ARE NATURAL OBJECTS composed of stone, metal or a mixture of metal and stone that survive their fall from space to land on the Earth. Most meteorites originate from asteroids, a few come from the planet Mars or the moon and some may even come from comets. Meteorites from asteroids are the oldest objects available for study, dating back to the very first stages of solar system formation, about 4,570 million years ago. The study of meteorites provides a window back in time, so we can learn about the processes that formed and shaped the solar system, and the planets, moons, asteroids and comets it contains, including our own Earth.

Meteorites are divided into three main types on the basis of their composition. Stone meteorites are composed mainly of silicate minerals similar to those found in rocks on Earth. Iron meteorites are composed of iron metal alloyed with nickel. Stony-iron meteorites, as their name suggests, are composed of a mixture of iron–nickel metal and silicate minerals. These different meteorite types are described further in Chapter 4.

## METEORS, METEOROIDS, FIREBALLS AND METEORITES

About 40,000–60,000 t of extraterrestrial material falls on the Earth every year. Before entering Earth's atmosphere these fragments are known as meteoroids, and individual pieces can range in size from tiny, microscopic grains of dust to large bodies, many metres in diameter. The vast majority of the material that lands on Earth is in the form of tiny dust grains, usually less than 1 mm ($\frac{1}{25}$ in) in size, known as cosmic dust, micrometeorites or interplanetary dust particles (IDPs). Because these objects are so small, they radiate away the frictional heating produced as they pass through the atmosphere faster than they melt, and thus survive intact to land on Earth.

You may have been lucky enough to have seen a meteor or 'shooting star'. These streaks of light are produced by larger objects, perhaps the size of a small pebble.

OPPOSITE Weighing in at 60 t plus, the Hoba meteorite is the largest known on Earth.

RIGHT The three main types of meteorite: stone, iron and stony-iron. Irons and stony-irons are the product of melting, as are stony achondrites. Chondrites have not melted since they first aggregated.

Stone

Chondrites — Unmelted

Achondrites

Iron

Melted

Stony-iron

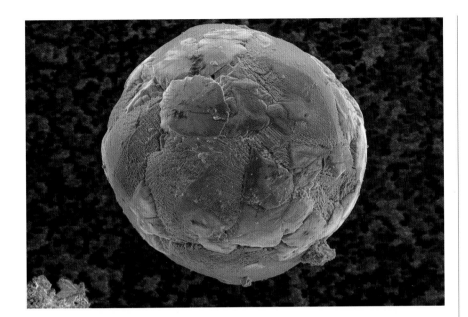

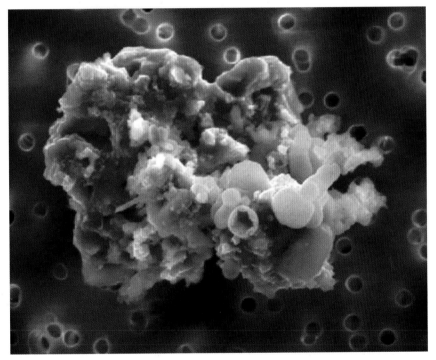

LEFT AND BELOW LEFT A cosmic spherule collected from the Antarctic ice (field of view is about 60 μm), and an interplanetary dust particle collected in the atmosphere (field of view is about 25 μm). Tens of thousands of tons of material like this fall to Earth each year.

The effects of frictional heating as such objects pass through Earth's atmosphere are great enough to melt and vaporize the material, producing a very rapid, bright streak in the night sky. At some times during the year many meteors can be observed over a period of a few days. These 'meteor showers' are related to the orbits of known comets and occur at well-defined times during the year (see Chapter 5). Meteors do not result in meteorites. It is extremely rare for material from a meteor to be recovered on Earth.

If still larger fragments (many centimetres or even metres in size) enter Earth's atmosphere they can produce spectacular light and sound effects as they fall. As they pass through the Earth's atmosphere these are called fireballs. Meteorites may land on Earth as a result of a large meteoroid entering the Earth's atmosphere. A meteoroid travels in space at its cosmic velocity, about 30 km/s (67,500 mph). When it reaches the Earth's atmosphere it is slowed down by friction and these frictional forces also act on its outermost surface, heating it to melting temperature.

RIGHT Scientists collecting an ice core in Antarctica in order to extract micrometeorites.

BELOW RIGHT The Peekskill meteorite was witnessed by many people as it partially burned up in the atmosphere.

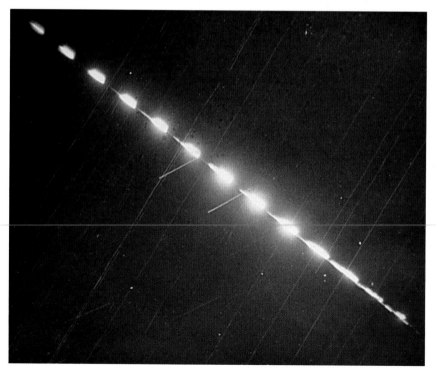

Only the outermost surface of the meteoroid melts and the resulting droplets of molten material are carried away as the meteoroid speeds through the atmosphere. (This is unlike meteors, where the entire object is vaporized and destroyed.) The rapid speed of the meteoroid as it enters the atmosphere also generates shock waves (sonic booms), which are often heard as explosions. As the meteoroid is further slowed down by the atmosphere it begins to cool rapidly and its molten surface forms a glassy coating called the fusion crust. Finally, it falls to the ground under the influence of gravity and is now termed a meteorite.

LEFT The fusion crust of a meteorite is a thin coating that may be glassy or matt in appearance.

## METEOROIDS, METEORS AND METEORITES

A meteoroid is a small body typically less than hundreds of kilogrammes in mass and about 1 m (3¼ ft) in diameter travelling through space, and which may (or may not) land on Earth as a meteorite. A meteor or 'shooting star' is a vivid streak of light produced when a small meteoroid, perhaps up to a few centimetres in size, enters the Earth's atmosphere and completely vaporizes through frictional heating. Meteorites are natural objects made of rock, metal or a mixture of rock and metal that survive their fall from space to land on the Earth.

Sometimes one side of the meteorite may be curved or even conical in shape. This is the 'leading edge' of the meteorite, the side that pointed in the direction the meteoroid travelled during its passage through the atmosphere. If the meteorite has tumbled during its flight it will have a more irregular shape, although it should still retain rounded edges. Note that only the outermost surface of the meteoroid melts, while the interior remains cold and unchanged. Meteorites are almost always cold when they land. The very few reports of meteorites, that have been recovered almost immediately after landing, state that they are cold or slightly warm at most.

Meteorites fall more or less randomly over the Earth's surface and a number of studies have been carried out to estimate how many meteorites land on Earth every year. Some studies use cameras to image the night sky and record the number of observed fireballs, while others use the number of meteorites found in desert regions. Estimates indicate that approximately 5,000–17,000 meteorites larger than 100 g (3½ oz) – about the size of a golf ball – fall on Earth every year. However, it is

BELOW The conical shape points in the direction of travel during atmospheric flight. The ridges that flow away from the centre were produced by ablation during entry.

important to remember that 70% of the Earth's surface is covered by water and any meteorites landing in the oceans are unlikely ever to be recovered. Similarly, many parts of the world remain uninhabited or only have sparsely distributed populations, so any meteorites landing in these areas are unlikely to be seen by anyone. Every year only a handful of meteorites are witnessed to fall and are then recovered – these are known as 'fall' meteorites. For example, there were six witnessed falls in 2014, eight in 2015 and eleven in 2016.

In some places, however, meteorites can be preserved or concentrated by natural processes. Most meteorites contain some proportion of metal, and the arid conditions found in the hot and cold deserts help to preserve them by slowing down the weathering and rusting that destroys them. Hot and cold deserts have provided extremely fertile meteorite hunting grounds, with tens of thousands of meteorites recovered so far. Meteorites that are found after they fell, either by chance or during dedicated collecting trips, are termed 'find' meteorites. Meteorite falls and finds are discussed further in Chapter 2.

ABOVE A fragment of the Sarıçiçek howardite meteorite, that fell in Turkey in September 2015. A fireball was seen on several security cameras and small pebbles of the meteorite rained down onto houses in the village of Sarıçiçek.

# THE STUDY OF METEORITES – FIREBALLS AND THUNDERSTONES

Meteors, fireballs and meteorite falls have been observed and recorded since ancient times. Chinese astronomers recorded meteor showers around 700 BC, and objects including daggers and jewellery made from meteoritic iron have been recovered from Egyptian pyramids, including the famous tomb of Tutankhamun. In 1925, two daggers were discovered alongside Tutankhamun's mummified body. One, wrapped against the right thigh, was found to contain nickel in the iron blade, which indicates that this iron is meteoritic in origin. In the thirteenth century BC the Egyptians used a hieroglyph which is translated as 'iron from the sky', indicating that they knew about the fall of iron meteorites.

In 1928, archaeologists excavating a prehistoric Native American settlement near the town of Winona in northern Arizona, USA, excavated a 24-kg (53-lb) meteorite that had been carefully buried in a purpose-built crypt, or 'cist'. The archaeological setting suggests that the builders of the settlement treated the meteorite as an important object. However, the Sinagua tribe was only in the area for around 200 years from the eleventh century, and therefore the meteorite is unlikely to have been a witnessed fall.

The oldest meteorite seen to fall, and which is still preserved, is the Nogata meteorite that fell in the garden of the Suga Jinja Shinto shrine in Japan, over 1,100 years ago. Realizing the importance of the object, the shrine priests preserved it in a box as a special treasure. Although no written records exist, the story of the fall was passed

RIGHT The Ensisheim meteorite, which fell in 1492. Its fall was recorded in drawings and engravings and the stone was preserved. The stone can now be seen in the town hall in Ensisheim, France .

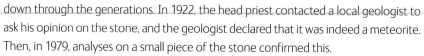

down through the generations. In 1922, the head priest contacted a local geologist to ask his opinion on the stone, and the geologist declared that it was indeed a meteorite. Then, in 1979, analyses on a small piece of the stone confirmed this.

The first witnessed meteorite fall for which written records exist is the Ensisheim meteorite, which fell on the village of Ensisheim (now in the Alsace region of France) in 1492. A large fireball was observed and, after a large explosion – which caused much fright and excitement among the local people – a 127-kg (280-lb) stone fell on Ensisheim. This event was deemed so unusual and important it was recorded at the time in an engraving and was reported to the Emperor Maximillian, who ordered that the stone be preserved in the village church. The stone can still be seen in Ensisheim although it now resides in the town hall.

Between the fall of the Ensisheim meteorite and the late 1700s a number of meteorites were seen to fall and were recovered, although they were considered to be little worthy of scientific study. The modern study and science of meteorites (termed meteoritics) really began in 1768 when a stone that was claimed 'fell from the sky' over the town of Lucé, France, was presented to the Abbé Bachelay. The Abbé collected a number of eyewitness reports of the stone's fall and recovery, and also wrote a detailed report, which he then passed on to the French Academy of Sciences. The Academy set up a commission of enquiry, which included the famous French scientist Antoine Lavoisier, to study the stone and the Abbé's report.

BELOW The German scientist, Ernst Chladni (1756–1827) was the first to propose that meteorites are extraterrestrial in origin and is one of the founders of the modern scientific study of meteorites.

The commission's members undertook the first chemical analyses of a meteorite. They recognized the presence of iron sulphide and concluded that the stone was simply a terrestrial pyrite (pyrite, chemical formula $FeS_2$, is also known as 'fool's gold'). To account for the eyewitness reports of its fall and the presence of a black fusion crust, the scientists stated that the stone must have been struck by lightning; consequently, meteorites were commonly referred to as 'thunderstones'. This report, by such an eminent group of scientists, received wide acceptance among the intellectual elite of Europe. Scientists were unwilling to believe in a phenomenon for which no witnesses of suitably high social rank could be found, and for which no reasonable scientific theories could be provided. Until the end of the eighteenth century meteorites remained to be seen by the scientific community as 'nothing out of the ordinary', and some people even advocated their destruction!

A number of well-witnessed meteorite falls occurred between 1789 and 1803, and these generated the

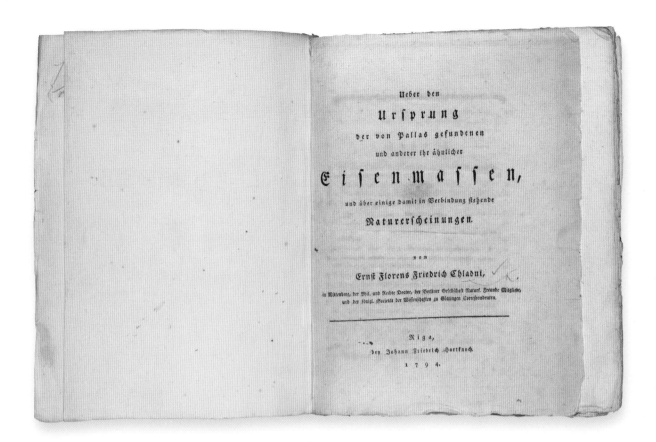

Ueber den

**Urfprung**

der von Pallas gefundenen

und anderer ihr ähnlicher

**Eisenmassen,**

und über einige damit in Verbindung stehende

**Naturerscheinungen.**

von

**Ernst Florens Friedrich Chladni,**

in Wittenberg, der Phil. und Rechte Doctor, der Berliner Gesellschaft Naturf. Freunde Mitgliede,
und der königl. Societät der Wissenschaften zu Göttingen Correspondenten.

Riga,
bey Johann Friedrich Hartknoch.
1 7 9 4.

publication of many books, articles and pamphlets to explain their nature and origin. In 1794, the German physicist Ernst Chladni published a short book on meteorites. In this book, he described the characteristics of a number of unusual iron masses and descriptions of meteors, fireballs and witnessed falls of stones and irons, and concluded that that they were extraterrestrial in origin.

The year 1794 also marked the fall of the Siena meteorite in Italy. This event generated much interest in the English expatriate community, the majority of whom came from the upper echelons of English society, and who were living in Italy or visiting it as part of the 'Grand Tour'. Travellers returning to England brought with them specimens of the meteorite (indeed, a thriving market grew up selling bogus 'fallen stones' to tourists), and fascinating stories regarding the shower of stones over the town.

Just a year later, in 1795, a large stone meteorite fell on farmland at Wold Cottage in Yorkshire, UK. The event was witnessed by several people, loud explosions were heard and a young ploughman, John Shipley, witnessed the stone fall through the clouds and onto the ground very close to him. The local landowner, Captain Edward Topham, obtained sworn eyewitness testimonies from three people who had witnessed the fall and from others who had heard the explosions. Early in 1796 Topham submitted a detailed account, which was published in the local newspaper, and he later took the stone to London where it was exhibited to the public. One

ABOVE The title page of Ernst Chladni's book *On the Origin of the Pallas Iron and Others Similar to it, and on Some Associated Natural Phenomena*, which was published in 1794.

of the visitors to the exhibition was the famous British naturalist and President of the Royal Society Sir Joseph Banks, and he obtained a sample of the stone. Banks decided that serious scientific investigation of meteorites should be carried out and so he passed his Wold Cottage fragment, along with samples of other meteorites, to the chemist Edward Howard. Howard discovered that all the samples contained nickel dissolved in iron, a characteristic that was not observed in any terrestrial rocks. Like Chladni, he concluded that an extraterrestrial origin was plausible.

The reports of Edward Howard's analyses caused many European chemists and mineralogists to reinvestigate samples reported to have fallen from the sky. If any remaining doubts were held by the scientific elite of Europe that the fall of meteorites was a real phenomenon, these were quashed when a spectacular meteorite shower of over 3,000 stones fell over L'Aigle, France, in 1803. However, where meteorites formed was still a matter of debate, with two main theories competing in the scientific discussions of the early to mid-nineteenth century. Some scientists believed that the stones formed in the upper parts of the Earth's atmosphere, while others believed that they were formed as the result of volcanic explosions on the moon. It is important to stress that at this time the Earth–moon system was considered to be 'closed' (i.e. nothing could be added or lost), and that all atmospheric and weather phenomena (clouds, rain, hail, meteors, fireballs, etc.) were part of this system. Chladni's theory that meteorites were truly extraterrestrial, originating outside the Earth–moon system in cosmic space, was still little believed.

Continued investigation of newly fallen meteorites in the early to mid-nineteenth century revealed that not all meteorites had the same chemistries or textures, or contained the same minerals, which suggested that they could not all originate from one place (i.e. the moon). During the same period, advances in telescope design and manufacture allowed astronomers to study the moon in ever-increasing detail, with the result that few astronomers believed that lunar volcanoes were still active. Added to this was the finding that meteorites entered the Earth's atmosphere at cosmic velocities and, therefore, at too high a velocity to be of lunar origin. Finally, since their initial discovery in 1801, an increasing number of asteroids was being observed; these were then thought to be the remains of a fragmented planet. It did not seem unreasonable to suggest that some of these fragments could land on Earth as meteorites, and in 1849 the German scientist Alexander von Humboldt even called meteorites 'the smallest of all asteroids'.

With meteorites firmly being established as coming from space, their precise origin was still a matter of debate from the mid-1800s until as recently as the 1950s (interestingly, the Martian origin of the Shergotty, Nakhla and Chassigny ('SNC') group of meteorites was only firmly settled in the 1980s, see p.94). An interstellar origin (i.e. coming from outside our solar system) for some meteorites was suggested. However, this theory was dismissed when precise measurement of velocities and orbits for meteorite parent bodies were obtained, as these showed they must have originated within the solar system.

Meteorites, rightly, are now recognized as being rare and precious samples of bodies within our solar system. They record a wealth of information about the birth and growth of the solar system; the formation and history of planets; the role of impacts in shaping and modifying planets, asteroids and moons; and they even record evidence of astrophysical processes that occurred long before our solar system was born.

CHAPTER 2

# Where are meteorites found?

VERY YEAR THOUSANDS OF METEORITES land on Earth. They have been recovered from every continent of the Earth. In some places, where the environment and geography are favourable to meteorites being preserved and concentrated, many thousands have been found. Meteorites that are witnessed to fall (often after people have observed spectacular fireballs and heard loud explosions), and are subsequently recovered, are known as meteorite 'falls'. As of January 2018, there have been 1,162 authenticated meteorite falls.

Meteorites that are found either accidentally or during a meteorite collecting trip are known as 'finds'. By January 2018, there were roughly 56,000 recorded find meteorites (this is an increase of approximately 20,000 since the previous edition of this book was published in 2009). Meteorites are usually named after the nearest town or geographical location where they fell or were found. In hot and cold desert regions, where many thousands of meteorites can be found in a relatively small area, they are given a name relating to the geographic region and a number (e.g. Allan Hills 84001, Dhofar 342).

OPPOSITE The arid climate and desert landscape of the Nullarbor Desert, Western Australia make it an ideal location to hunt for meteorites.

LEFT Part of the Barwell meteorite, which fell on Christmas Eve 1965, smashed through a window as it landed.

## METEORITE FALLS

The rate of meteorite falls can be calculated using observations of fireballs and meteors. Data show that meteorites fall almost evenly around the globe, although they are slightly less likely to fall at the poles than at the equator, because the parent meteoroids are likely to be travelling in the same plane as the planets. Meteorites can fall at any time of the day, but observations suggest that more meteorites fall in the afternoon and evening. This may be because at this point in its rotation, the Earth is facing in the same direction as it is travelling around the sun; therefore, meteorites are much more likely to fall (much in the same way as flies are more likely to stick to the front window of a car as it is travelling forwards).

Before the mid-twentieth century, the statistics of meteorite falls relied solely on eyewitness reports. Although these reports are often extremely detailed and provided by reliable witnesses, it must be remembered that a number of human and social factors can skew the data. Perhaps the most obvious is that countries and regions with the most recorded meteorite falls are also those with the highest population densities (e.g. Europe and Japan); the more people there are to see a meteorite fall, the higher the chance of recovering it. The type of meteorite can also play a part. Iron meteorites are more likely to be recognized by non-experts and thus are more common among 'find' meteorites. 'Fall' meteorites are dominated by stony meteorites.

Fireballs are much easier to see at night; however, it is much easier to see any associated meteorite land if it is daylight. People may be outside, working or travelling, during the hours of daylight, and therefore there are more potential witnesses to any meteorites falling to the ground. Similarly, seasonal variations in

BELOW World map showing the location of where meteorites have been seen to fall. Note that more meteorites are seen to fall in high population regions, places where there are more people to see a meteorite fall.

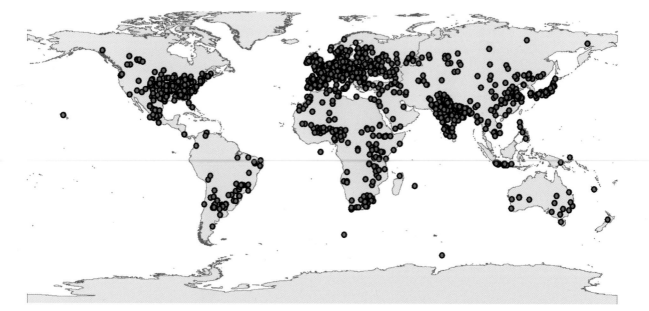

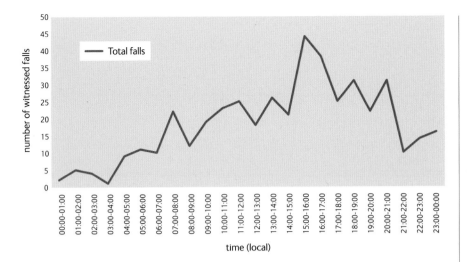

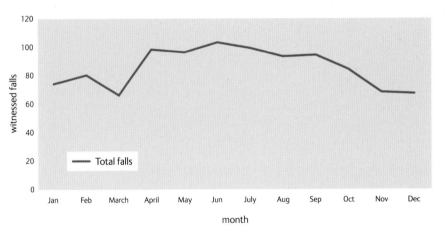

LEFT Graphs showing recorded time of day (upper) and dates of witnessed meteorite falls (lower).

recorded meteorite falls could be explained by the ease (or difficulty) of finding and recovering meteorites. It may be easier to find a meteorite on a frozen lake or in a barren winter field, compared with a highly vegetated area. However, it is likely that more people will be outdoors during spring and summer, so more witnesses are around not only to see a fireball but also to hunt for any potential meteorites. Data seem to indicate that the spring and summer months are advantageous for witnessing meteorite falls.

In an effort to better understand the relationship between meteors, fireballs and meteorites, camera networks dedicated to observing the night sky have been in operation since the 1950s. These cameras have recorded many thousands of meteors and fireballs streaking through the atmosphere and, in a handful of cases, have provided scientists with proof that a meteorite has landed and, more importantly, where it has landed.

In addition to providing much more accurate data on how much extraterrestrial material lands on Earth each year, these camera observations have also provided some very important information in our study of meteorites and their place in the

ABOVE A camera station of the Desert Fireball Network located in the Nullarbor Desert of Western Australia.

solar system. We now have a better understanding of the numbers of meteorites that are arriving on Earth each year. The velocities and trajectories of fireballs have provided scientists with definitive proof that meteorites originate in the solar system. Finally, if the fireball is large enough and if there are sufficient data, it is possible to determine the potential size of any resultant meteorite(s) and also to determine with a fair degree of precision where it – or they – will have landed.

Unfortunately, there are only a few camera networks currently in operation around the world that are dedicated to recording fireballs and any resultant meteorite falls: the 'Desert Fireball Network' in Western and South Australia; the 'Spanish Meteor Network' and the 'European Fireball Network' in central Europe; and the 'Fireball Recovery and Interplanetary Observation Network' in France. Since its start in 2007, the Desert Fireball Network has observed numerous fireballs and has allowed for four meteorites to be recovered. The first success came just after the network began operating in its testing phase. In 2007 a large fireball was recorded by cameras situated in the Nullarbor Desert. Not only did calculations indicate that a meteorite had landed, but it was also possible to determine where the meteorite would be. In late 2008, a search team visited the area to hunt for the meteorite. Two fragments weighing a total of 324 g (11½ oz) were found, only 100 m (328 ft) from the predicted landing site!

Occasionally, large fireballs are witnessed by enough people to enable a large number of accurate eyewitness accounts to be gathered. With security and CCTV cameras becoming more and more common, fireballs are often also recorded on video. The eyewitness accounts and associated video camera footage may be sufficient to calculate the fireball's velocity and trajectory, and to determine the potential landing site of any resultant meteorite(s). A famous example of a very well-observed and well-recorded fireball is the Chelyabinsk 'event' of February 2013 in Russia. This spectacular fireball event is strictly termed a 'superbolide' as the fireball was so bright – at its brightest it was the same brightness as the sun. This superbolide was recorded by numerous security cameras and 'dash cams' in cars. The video evidence proved extremely useful in studying this spectacular and unexpected event and was used by scientists to estimate the speed, size and origin of the asteroid that caused the superbolide (see also p.38).

BELOW A large fireball was observed from several places in the UK on 17 March 2016, at 3:16am. Images like this help scientists to determine the original orbit of the meteorite before it came to Earth.

BELOW Map showing the locations of eyewitnesses to the large fireball witnessed over a number of northern US states and Canada on the night of 16 January 2018. The American Meteor Society website has a fireball reporting website (http://www.amsmeteors.org/members/imo/report_intro) and for this event 674 people reported what they saw. This data proved very important in aiding in the discovery of the meteorites that landed.

A recent advance in detecting the fall of a meteorite comes from analyzing weather radar. Once the fireball is slowed down by the Earth's atmosphere and loses its cosmic kinetic energy, it ceases to be luminous and the meteorite enters a period known as 'dark flight', which typically occurs at a height of approximately 25 km (15½ miles – an airliner flies at approximately 11 km (almost 7 miles). During this time the meteorite will be falling under the influence of gravity, and its trajectory and subsequent landing site can be heavily influenced by strong winds in the atmosphere. Weather radar can be used to observe falling meteorites up to heights of 10 km (6¼ miles) and can provide models of both the mass and the location of the meteorites that have landed. This technique has been used to aid in the recovery of a number of meteorites in the USA, the most recent of which was in January 2018.

## KNOWN METEORITES AS OF JANUARY 2018

|  | Falls | Finds | Total | Falls as % of total |
|---|---|---|---|---|
| Stones | 1,103 | 55,147 | 56,250 | 1.96 |
| Stony-irons | 11 | 339 | 350 | 3.14 |
| Irons | 48 | 1,133 | 1,181 | 4.06 |
| Total | 1,162 | 56,619 | 57,781 | 2.01 |

# RECOVERING METEORITES WITH RADAR TECHNOLOGY

The two images show composite views of weather radar signatures of falling meteorites from the Hamburg, Michigan, USA meteorite fall at approximately 8pm on 16 January 2018 (local time). The grey/blue polygons are radar reflections off falling meteorites, and the two images show the same data viewed from along (right) and across (below right) the long axis of the fall. These data are taken from a total of six radar sweeps made by two different radars. The first signature appears 269 sec after the end of the fireball, to account for the time when the meteorites fell from the fireball's altitude down to the radar's detection volume, and the last signature occurred 819 sec after the fireball. The radar signatures seen here range from 3,230 m (10,597 ft) altitude down to only 110 m (361 ft) above the ground. Radar images of falling meteorites are very accurate in terms of their determining their location on the ground, which allows for accurate identification of meteorite falls and rapid recovery of fresh, relatively unweathered meteorites. This data was used by meteorite hunters to locate the likely falling site of the meteorites and the first fragment was found on the morning of 18 January 2018. Since the 1980s when the radar technology appropriate for detecting meteorite falls was deployed over the USA and Canada, nearly 40 falls have been detected using radar and a number of meteorites have been successfully recovered.

**ABOVE** Images showing radar data of falling meteorites viewed from along (top) and across (above) the long axis of the fall from the large fireball witnessed across several northern US states and Canada on the night of 16 January 2018.

**LEFT** Image from a security camera of the large fireball witnessed over a number of US states on the night of 16 January 2018.

## OBSERVING AND RECORDING FIREBALLS

Although thousands of meteorites land each year, it is rare to witness a fireball and extremely rare to see a meteorite fall. Reports of fireball observations can be submitted to meteor and fireball sections of local and/or national astronomical societies. There are also a number of online tools to report fireball and meteor observations. A local university with an astronomy department may also be able to provide you with contact information of where to report your observations. If you are privileged to witness such an exciting and exceptional occurrence, there are certain observations you can record that will greatly help scientists investigate the event:

1. Record the time of day you saw the fireball/ meteorite fall.
2. If you saw a fireball, how long did it last?
3. How bright was the fireball (compared with the moon, stars or sun); was it coloured; did it have a form or shape?
4. Did the fireball/meteorite leave any form of trail? If so, what did it look like and how long did it last?
5. What was the direction of the fireball's travel through the sky? This should be related to fixed objects on Earth (e.g. tall buildings) or to star and planet positions in the sky.
6. Did you hear any sound effects? If so, what were they like and when did you hear them?

If you are one of the very rare people fortunate enough to see a meteorite actually land, there are some important things you can record and do:

1. How long after the fall was the meteorite(s) recovered?
2. Where was the meteorite(s) found? If possible GPS coordinates should be recorded and/or the location marked on a map.
3. How much did the meteorite(s) weigh?
4. What was the ground like where the meteorite(s) landed? Did the meteorite(s) penetrate the ground? If the meteorite(s) penetrated the ground, how deep was the hole?

Photographs and drawings of fireballs and any resulting meteorite falls add a huge amount of information and can be extremely useful. Remember if you are entering private property it is vital to gain the landowner's permission. It is also recommended that you do not touch the meteorite with your bare hands. Meteorites are not dangerous when you pick them up, but they can easily become contaminated with Earth material, which is detrimental to scientific study. Wear clean gloves, such as latex or polythene gloves and wrap the meteorite in clean aluminium foil and/or put it in a clean sandwich bag. A suspected meteorite fall should be reported to the nearest large natural history museum, science museum or university geology department. You can find out more about meteorite falls, including examples of 'close encounters' with them, in Chapter 3.

# METEORITE FINDS

Up to the mid-1960s only about 1,800 meteorites were known, and of these approximately 43% were witnessed falls. As of January 2018, roughly 57,000 meteorites are known, of which only 2% are witnessed falls (see table on p.24). The much greater numbers of meteorites being found today results from the fact that it is now known that certain of the world's desert regions are very fertile places to find meteorites. Every year, dedicated collection trips are organized by scientific parties to hot and cold deserts to hunt for meteorites.

## ANTARCTICA

The first meteorite recorded to be found in Antarctica was in 1912, during the Australasian Antarctic expedition led by Douglas Mawson. Over the next five decades only a handful more meteorites were discovered by various expeditions.

It was not thought to be anything unusual to find a few meteorites on this continent; meteorites fall all over the world so why wouldn't they be found in Antarctica?

In 1969, a group of Japanese scientists studying glaciers in Antarctica found nine meteorites near the Yamato Mountains. At first it was thought that these meteorites must have been fragments from a single fall. However, it was soon discovered that they comprised six different meteorite types and, therefore, could not have been from a single meteorite. Continued exploration in the mid- to late 1970s by Japanese teams resulted in many hundreds of meteorites being recovered from this area. After hearing about the amazing meteorite discoveries made by the Japanese, American scientists also started to make dedicated expeditions to Antarctica to hunt for meteorites. The first US expedition in 1976 to the Allan Hills area recovered 11 different meteorites of five different types. Today, in addition to the expeditions run by Japan and the USA, teams of scientists from Europe, China and South Korea all make regular trips to the Antarctic continent to search for meteorites. Up to January 2018, these collecting trips have resulted in over 36,000 meteorites, more than from any other place in the world.

BELOW 'Blue ice' field in Antarctica. These are regions where underlying ice is exposed because of the action of the strong winds. The surface of such regions can contain a very high density of meteorites.

RIGHT A mountain range called the Trans-Antarctic mountains runs across the Antarctic continent and there are areas around these mountains that are particularly good for meteorite hunting.

So, why is Antarctica such a good place to find meteorites? First, not only is the Antarctic plateau the coldest place on Earth, it is also the driest place. Any meteorites that land in Antarctica are very likely to be preserved in the dry, deep-freeze conditions, for hundreds of thousands or even millions of years. Second, meteorites in Antarctica can be very easy to spot. In areas where there are no other sources of rocks, the only dark objects to be found on the ice are likely to be meteorites. Finally, the movement of the ice concentrates meteorites in some locations. As the ice moves away from the south pole towards the oceans its movement becomes blocked by mountains (such as the Transantarctic Mountain range), forcing it into confined areas or 'cul de sacs'. Here the strong Antarctic winds (known as katabatic winds) erode or ablate ice from the surface, leaving a residue of rocks that were carried within it. This process concentrates any rocks (including meteorites) that are carried in the ice. The regions where ablation occurs are known as 'blue-ice', as the ice has a pale, bluish tinge. Satellite imaging can detect these blue-ice regions and so provide scientists with new areas in which to hunt for meteorites. Under the Antarctic Treaty signed in 1959, and which came into force in 1961, exploration in Antarctica cannot be used for mining and financial gain. All geological samples from the continent can only be used for scientific study. Therefore, all Antarctic meteorites recovered after the Treaty was signed are kept within scientific institutions and cannot be bought or sold.

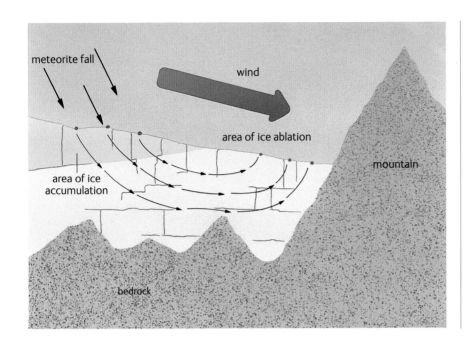

meteorite fall

wind

area of ice ablation

area of ice accumulation

mountain

bedrock

LEFT Diagram to show how meteorites may be concentrated in certain parts of Antarctica. As glaciers move towards the ocean they carry the meteorites with them. But if their path is blocked, the ice is stripped away by the wind exposing the residue of meteorites.

BELOW A meteorite on the ice in Antarctica. The dark meteorites are easy to spot against the icy background.

## HOT DESERTS

Hot desert areas are also favourable places for meteorite preservation, although unlike Antarctica there are no natural processes involved that concentrate the meteorites in specific areas. As of January 2018, more than 14,000 authenticated meteorites have been recovered from hot desert regions around the world. Parts of the Sahara Desert in Libya and the deserts of Morocco, Algeria and Oman have proved particularly fruitful for meteorite searches. Meteorites have also been recovered from other desert regions around the world including the Nullarbor Plain in the south of Western Australia, the Namib Desert in southwestern Africa and the Atacama Desert in Chile.

The very dry conditions in the hot deserts limit the amount of weathering and rusting of the meteorites and so enhance their preservation. The best places to look for meteorites in the hot desert are areas where the ground is flat, where there are few or no dark local rocks that could be mistaken for meteorites and where there are no sand dunes to cover up any meteorites. However, perhaps most importantly,

BELOW A 294 g (10 oz) partially fusion crusted stone from the Camel Donga eucrite strewn field. The orange/brown staining on the black, fusion crusted surface is staining from the local soil.

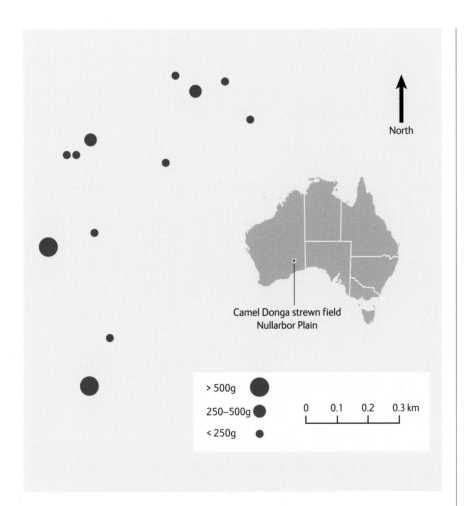

LEFT Map showing sites where stones of the Camel Donga meteorite have been found.

North

Camel Donga strewn field
Nullarbor Plain

> 500g

250–500g

< 250g

0    0.1    0.2    0.3 km

the area must have remained arid and unchanged for many thousands of years so that the number of meteorites 'builds up' over time. For example, the Nullarbor Plain has been a desert for over 30,000 years.

Hot desert meteorites are found in the place where they landed, and even in a particularly fruitful area you would be lucky to find two or three meteorites in a square kilometre. Of course, some areas have abundant meteorites where tens or even hundreds of samples may be found in a relatively small area of a few square kilometres. However, these 'strewn fields' are caused by a single meteorite breaking up during flight and producing a shower of stones over a limited area. A well-known site in the Nullarbor Plain is the Camel Donga strewn field, where several hundred fragments (weighing over 25 kg or 55 lb) of a eucrite meteorite have been collected.

In contrast to Antarctica, it is sometimes legal to buy or sell meteorites from hot deserts, or indeed from some other locations where they may have been found. However, some countries like Oman and Australia do have strict laws that restrict the commercial exploitation and export of meteorites. If you are interested in purchasing a meteorite you should make sure that the meteorite has not been illegally exported from its country of origin.

# CHAPTER 3

# Impacts and collisions

ALONG WITH HEATING AND MELTING, impact and collision are fundamental processes in the evolution of the solar system from its formation to the present. Collision is responsible for several effects in the solar system. Impacts at low energies (which mainly took place early on) caused dust-sized particles to stick together to form bigger and bigger bodies like asteroids and planetesimals and, ultimately, planets. As these small bodies collided and bounced into and off each other, the bits and pieces left over started speeding up and soon were causing much more destructive collisions. Impacts at higher energies have altered planetary surfaces such as forming the moon and craters, and causing some planets such as Venus to tilt on their axis. Impacts continue to occur today, although the frequency is much less. In this chapter, we discuss the evidence of impacts, how they have affected the solar system in the past and the hazard they present today.

## EVIDENCE OF IMPACTS

The most obvious effect of impact is a crater. The evidence for impacts in the solar system is not difficult to find. Just look at any of the rocky, inner solar system planets of Mercury, Venus, Earth and Mars, and at the moon, for confirmation of a very violent history. It is easiest to see the effects on planets that have not been geologically active for billions of years. Mars and Mercury, and also the moon, reveal evidence that these bodies were bombarded quite heavily over time.

The Earth has also been bombarded throughout its history, but geological processes such as volcanism, and weathering processes, have erased or obscured many craters. Currently, there are 190 well-characterized meteorite craters scattered across the Earth's surface. Most are found on the oldest, continental regions (called cratons) that have been stable for billions of years. The largest craters – at Vredefort in South Africa (300 km (186 miles) in diameter) and Sudbury in Canada (250 km (155 miles) across) – are also the oldest, at about 2,000 million years and 1,850 million years, respectively. There is no correlation between age and

OPPOSITE Barringer (or Meteor) Crater in Arizona, USA, is approximately 1.2 km (¾ mile) across and was produced by an impactor about 50 m (164 ft) wide. The remains of the impactor are known as the Canyon Diablo iron meteorite.

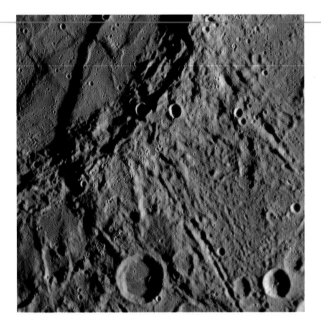

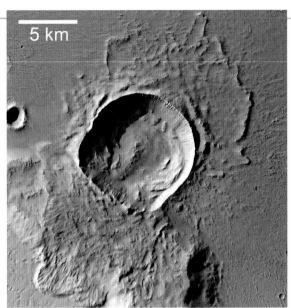

ABOVE LEFT These craters are on Mercury. The field of view is approximately 160 km (100 miles).

ABOVE RIGHT This crater on Mars has formed on a flat surface, and has expanded in a circular fashion. Because of the topography of the impact site, the shape of the crater was disturbed by the walls of the plateau and resulted in an asymmetrical shape.

RIGHT Impact craters on the moon; Glushko crater has a diameter of 43 km (26 ¾ miles).

crater diameter; for example, the Chicxulub crater in the Yucatan is 180–200 km (112–124 miles) in diameter and is 66 million years old. Older craters like these are often extensively eroded and are identified through recognition of shock-produced features in the bedrock (see opposite).

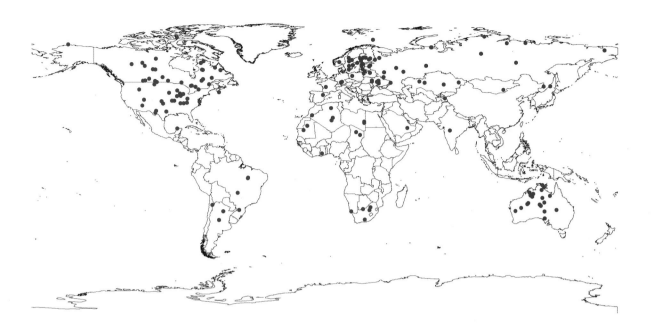

## CRATERS

Fast-moving meteorites form craters by impacting the Earth's surface so hard that they explode and excavate material. The 50,000-year-old Barringer Crater, also known as Meteor Crater, in Arizona, which is over 1 km (½ mile) across and about 200 m (656 ft) deep, was made by an iron meteorite approximately 50 m (164 ft) in diameter. An object of this diameter would weigh about 300,000 t. Over 30 t of an iron meteorite, known as Canyon Diablo, have been collected around Barringer Crater, accounting for approximately 0.01% of the mass of the impactor. The remaining material vaporized on impact. Some of this material is now found as small, metallic spherules in the soil around the crater. Fortunately, impacts of this magnitude occur only every 20,000 years or so, but, with the spread and speed of growth of population centres, and increasing urbanization of the landscape, humanity is more and more likely to suffer a catastrophic cosmic impact.

The mechanism behind crater formation is the reaction of rock to compression and decompression during shock. At impact, a shock wave moves forwards into the rock that has been hit and also backwards into the impactor, melting and vaporizing it. The high shock pressures compress the rocks. A decompression front follows the pressure wave that allows the shocked material to be ejected outwards in a sheet of material. The walls of the newly forming crater are lined with material vaporized from the impactor and the impact site. The floor of the crater consists of broken up and glassy material, rocks that have been distorted or melted during impact. They contain remnants of the impactor.

Simple craters (of which Meteor Crater is an excellent example), are less than 3–4 km (1¾–2½ miles) across. They are bowl-shaped, almost always circular depressions and are formed by impactors less than 1 km (⅔ mile) across. Larger impactors produce

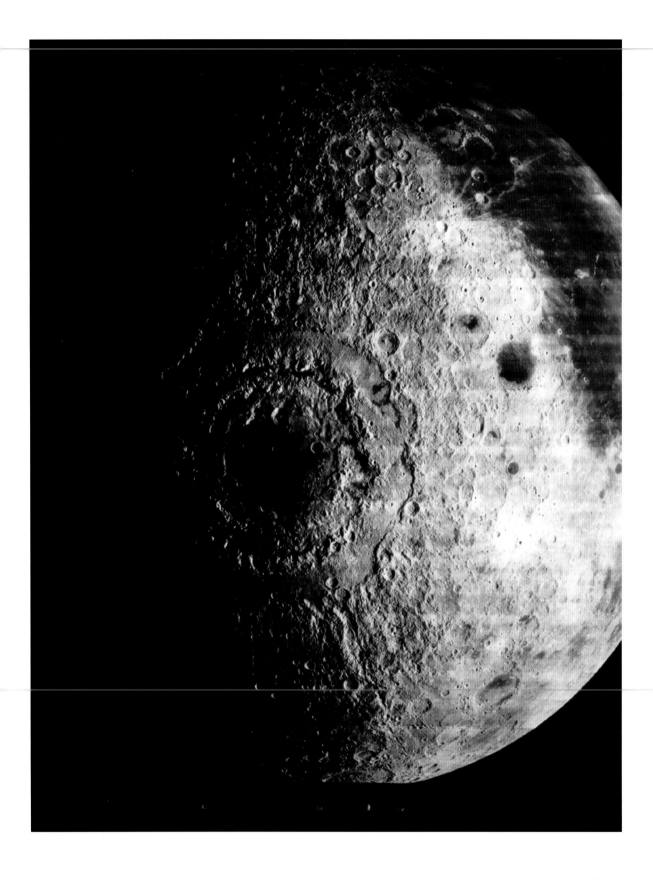

more complex craters, many kilometres across. Complex craters may have a central uplift, or plateau. This is produced as the shocked, unexcavated strata rebound from the impact, similar to water droplets on a pond rebounding and producing an initially raised central point from which ripples radiate. They are sometimes multi-ringed, the rings caused by slumping of the crater walls. The crater floor is often filled with a thick layer of shock-melted glass, or impact melt, which contains trace remnants of the impactor. Some of the most spectacular complex craters, or multi-ringed basins, are seen on the moon.

Not all meteorite impacts with Earth generate craters. Sometimes, a meteorite might have enough mass, but it breaks into smaller bits as it burns up in the atmosphere. These smaller pieces reach terminal velocity (also known as free-fall velocity) more rapidly, and these meteorites will 'bump' into the Earth with enough force to leave a dent, but not enough energy to form the characteristic features of a crater such as shock-produced minerals or overturning of materials. In 2007, such a meteorite crashed into the Earth in Peru, leaving a hole about 15 m (49 ft) in diameter and 5 m (16½ ft) in depth.

OPPOSITE The Mare Orientale basin on the moon, taken by Luna Orbiter 4. The multi-ringed basin is about 300 km (186 miles) across and was formed by a giant impact early in the moon's history.

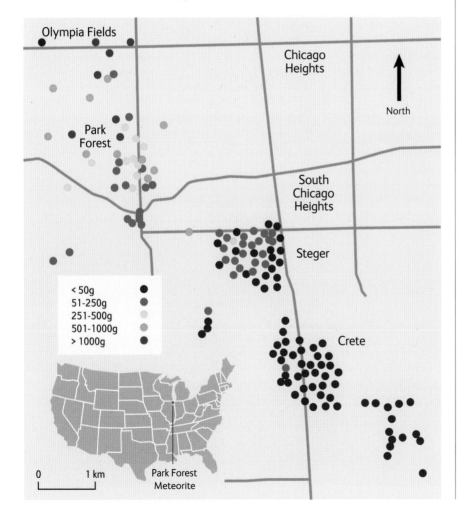

Olympia Fields
Chicago Heights
North
Park Forest
South Chicago Heights
Steger

< 50g
51–250g
251–500g
501–1000g
> 1000g

Crete

0    1 km
Park Forest Meteorite

LEFT A spectacular example of a meteorite that broke up in the atmosphere is the Park Forest meteorite that fell near Chicago, Illinois in 2003. The meteorite was estimated to have weighed thousands of kilograms before entering Earth's atmosphere, but broke up into smaller pieces the largest of which was just over 5 kg (11 lb). This map shows the locations where fragments of the meteorite were found.

## THE CHELYABINSK METEORITE

A spectacular example of an asteroid that broke up in the atmosphere and resulted in the fall of many hundreds of kilogrammes of meteorites is the Chelyabinsk meteorite that fell near Chelyabinsk, in the Chelyabinsk Oblast region of Russia, in February 2013.

The parent asteroid is estimated to have been approximately 20 m (65½ ft) in diameter and weighing in at around 13000 t before entering the Earth's atmosphere, but it broke up into thousands of smaller pieces, the largest of which weighed 654 kg (1442 lb) and was recovered from sediments at the bottom of Lake Chebarkul 6 months after the meteorite fell.

The asteroid entered Earth's atmosphere at 09:20 (local time, GMT +6:00), travelling at around 19 km/s (almost 12 miles/s). This large object, travelling at very high speeds, resulted in a large shock wave with the asteroid beginning to break up at a height of about 83 km (51½ miles) in the atmosphere, with a final fragmentation at around 27 km (16¾ miles). The extensive video and camera footage from the area allowed scientists to calculate that the energy of the shock wave produced during this event was in the region of 590 kT (the atomic bomb that exploded over the city of Hiroshima in August 1945 released an energy of about 15 kT).

More than 1,600 people sought medical attention and 112 were hospitalized for treatment, the majority of the injuries being caused by flying glass. Eyewitnesses also reported eye pain if they had looked directly at the fireball and many also reported feeling heat or warmth. Many buildings in the area were damaged, with the vast majority having windows blown in or out.

ABOVE This fragment of the Chelyabinsk meteorite shows a very shocked texture, with dark melt veins and fractures. It was found only four days after the fall, and has a fresh fusion crust where the outer material melted from heating in the atmosphere. The specimen is approximately 4 cm (1 ½ in) across.

BELOW Image showing the spectacular dust trail remaining following the atmospheric passage of the Chelyabinsk superbolide. So much dust was produced from this event it was detected by a NASA satellite and four days later the dust had travelled all around the world.

# THE TUNGUSKA IMPACT EVENT

ABOVE Part of the devastated region of Tunguska in Siberia photographed during the 1927 expedition to the area.

Early in the morning of 30 June 1908, a very bright fireball was visible over the remote Tunguska region of Siberia. Contemporary reports record that explosions like thunder were heard. In Russia, Germany and the UK, seismographs recorded tremors and barographs recorded air pressure waves at the time of the explosions. It was not until almost 20 years later, in 1927, that the Russian scientist Leonid Kulik led an expedition to the region to search for the remnants of what was presumed to have been a meteorite impact. Kulik did not find a meteorite crater, but an area of devastation about 60 km (37 miles) across. A belt of dead trees that appeared to have been blasted in an intense forest fire surrounded an approximately circular area of swamp. Alignment of the trees, pointing radially outward from the swamp, suggested that they had been felled by a shock wave. Only small particles of meteoritic material have been found in peat and tree resin in the area. It is assumed that an impactor (possibly 50 m (164 ft) across) exploded 6–8 km (3¾–5 miles) up in the atmosphere and the heat and shock wave from the explosion burnt and blew down the trees. It is still not certain whether the impactor was a comet or an asteroid. The energy of the impact has been estimated as 1017 J (equivalent to 50 million t of TNT).

## SHOCK FEATURES

Recognition of craters is often very easy: the bowl-shaped depression of Barringer Crater in Arizona is unmistakable. However, this crater is, geologically speaking, a very young crater in a semi-desert landscape. Craters that formed millions of years ago might be buried by sediments, eroded or otherwise obscured or erased. The shape of the landscape is therefore usually insufficient to allow recognition of a crater, although aerial photography can be used to find rounded features that can be investigated in more detail.

There are features preserved in impacted rocks that record the impact shock. These include microscopic signs, such as the presence of different forms of a mineral (e.g. stishovite and coesite, which are shock-produced forms of quartz), or deformation features within a mineral. The occurrence of microtektites (melted beads of glass) and tiny diamonds, formed from carbon as the result of the intense shock pressures, also indicate shock processing.

BELOW Aerial view of the Pingualuit Crater in Quebec, Canada, which formed approximately 1.4 million years ago.

LEFT Magnified view of quartz crystals showing deformation features. The field of view is about 0.5 mm (¹⁄₅₀ in) across.

BELOW LEFT Stereomicroscopic view of microtektites (the three pale-yellow, transparent glass spherules among the host detritus) collected from the tops of various Victoria Land Transantarctic Mountains, Antarctica.

0.5 mm

Another feature that can be used to determine an impact origin is the presence of shatter cones. These are cone-shaped fragments of rock that form as a result of the high velocity and pressure of the shock wave produced by the impact. They can be seen in rock exposures that might have been several kilometres away from the impact origin. The apex of the shatter cone always points to the centre of the impact, and thus mapping the distribution and orientation of shatter cones can help define an impact site.

BELOW Shatter cones in rocks associated with the Vredefort impact structure in South Africa. The pen is about 12.7 cm (5 in) long.

## IMPACT RATES

The amount of extraterrestrial material that falls on the Earth each year has been estimated at between 40,000 and 60,000 t. Almost all of this material arrives as dust grains much less than 1 mm (1/25 in) across – about two particles per square kilometre per second hit the Earth. Larger bodies impact less frequently; fewer than 40 meteorites weighing over 100 kg (220½ lb) fall each year. Although this may seem like a large mass, it takes an object weighing a minimum of 50,000 kg (110,231 lb) to make a crater. Because much of the Earth's surface is covered by water, or is uninhabited, most of these meteorites fall unobserved. Only about a dozen are seen to fall each year.

One way of calculating a cratering rate for the Earth is to use the number and ages of craters on the moon. Since the moon is so close to the Earth, the cratering rate of Earth and moon is likely to have been similar. But the moon is a geologically inactive, airless body (it has no atmosphere to protect it from impactors by slowing them down), and so craters on its surface have been less modified through time than those on the Earth. Based on lunar craters and terrestrial continental craters,

BELOW A stamp issued in 1957 to commemorate the fall of the Sikhote-Alin meteorite in 1947.

the rate at which craters with diameters greater than 10 km (6 miles) are formed on Earth is estimated at less than four craters per million years. The rate is thought to be lower now than in the earliest epoch of Earth's history, when the inner solar system had not been fully cleared of the debris remaining from planetary formation.

As the size of a meteorite increases, the smaller the number that fall each year, but the greater the potential impact hazard. Based on records of meteorite falls over the last 200 years combined with theoretical modelling results, two meteorites weighing about 1,000 kg (2205 lb) are expected to fall roughly every year, and one weighing about 100,000 kg or 1 t (220,462 lb) every 100 years or so. The biggest meteorite to fall in recent times landed in thick forest in the Sikhote-Alin mountains in Russia, in February 1947. Approximately 23 t of material was recovered at the time and at least a further 4 t has been unearthed subsequently. The fall was accompanied by a spectacular shower of fireballs, which made over 100 impact holes in the ground, the largest of which was almost 30 m (98½ ft) in diameter.

# HAZARDS FROM IMPACTS AND COLLISIONS

The impact hazard from the fall of a meteorite in the 1–10 kg (2¼–22 lb) size range is minimal. A person receiving a direct hit would probably die, and yet there are no well-documented instances of human fatality from a local meteorite impact. The most widely reported injuries associated with impact are those of a young boy in Mbale, Uganda, and a woman in Sylacauga, Alabama, USA. In August 1992, the boy was playing football with his friends, when a shower of about 50 stones fell. The stones ranged in size from 0.1 g to 27.4 kg (³/₁₀₀–60½ lb), and the boy was hit on the

RIGHT This young boy was lucky to escape serious injury when the Mbale meteorite fell in 1992.

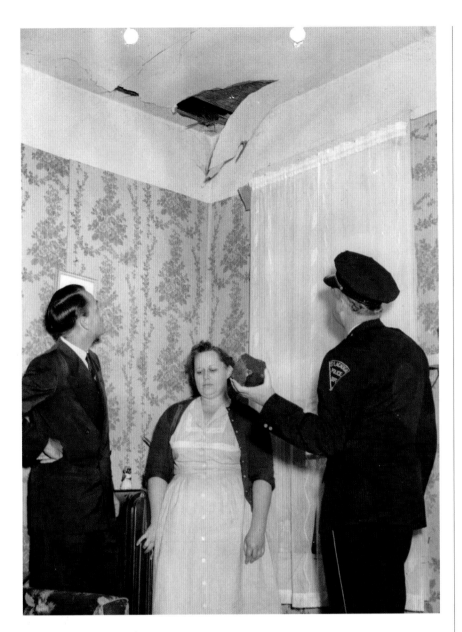

LEFT In 1954, Ann Hodges was hit on the side by a meteorite that came through the roof of her house while she was asleep.

head by one of the smaller stones, weighing approximately 3 g ($^1/_{10}$ oz). He escaped with bruises rather than any more serious injuries because the fall of the stone had been broken by the leaves of a tree. In November 1954, a lady called Mrs Ann Hodges was resting in her home in Sylacauga, Alabama, USA, when a piece of 5 kg (11 lb) meteorite crashed through the roof of her house, hit a radio and then struck her on the hip and her hand. Initially Mrs Hodges thought a gas heater in the room had exploded but then noticed a rock lying on the floor and a hole in the ceiling of her living room. Mrs Hodges was the first known person to bit hit by a meteorite and it caused a media sensation. Whilst she suffered bruising and swelling to her hip and hand she sustained no other physical injuries.

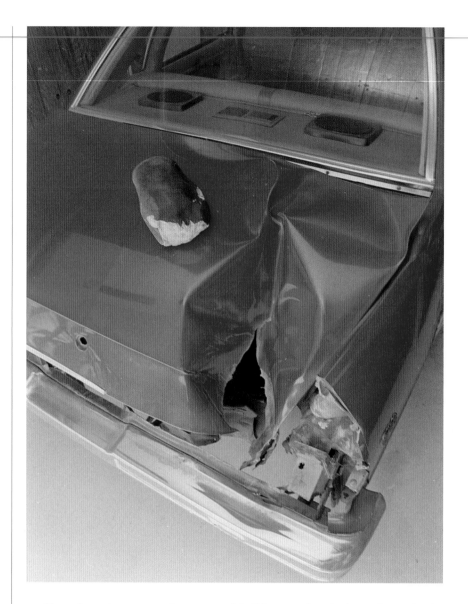

There have been many instances of buildings and parked cars being hit by meteorites. In 1938 the meteorite that fell in Benld, Illinois, USA, managed to hit both a building and a car as it fell through the roof of a garage and into the car parked inside. After passing through the floor of the car, the meteorite bounced back and finally embedded itself in the back seat. More recently, in the early evening of 9 October 1992, a stone meteorite fell onto a Chevrolet parked in Peekskill, New York, USA. The Peekskill meteoroid was accompanied by a bright fireball (see p.8) that was observed and recorded as it travelled across the northeastern USA, before landing in the boot (or trunk) of the car (see above). The main hazard to the Earth is posed by the so-called near earth objects (NEOs). These are asteroids and comets that have orbits that bring them to within 45,000,000 km (27,961,704 miles) of the Earth's orbit.

Most asteroids have very well-defined orbits that fall between the orbits of Mars and Jupiter in the asteroid belt, while comets (see Chapter 5) have orbits much further from the sun. Sometimes the orbits of these small bodies can be nudged by the massive planets and they can head inwards towards the Earth. Most of the NEOs are asteroids, rather than comets. The asteroids are called near earth asteroids, and they are divided into three groups called the Amor, Aten and Apollo asteroids.

BELOW Diagram of the inner solar system illustrating the orbits of the main belt asteroids and the near Earth asteroid groups – the Amors, Atens and Apollos.

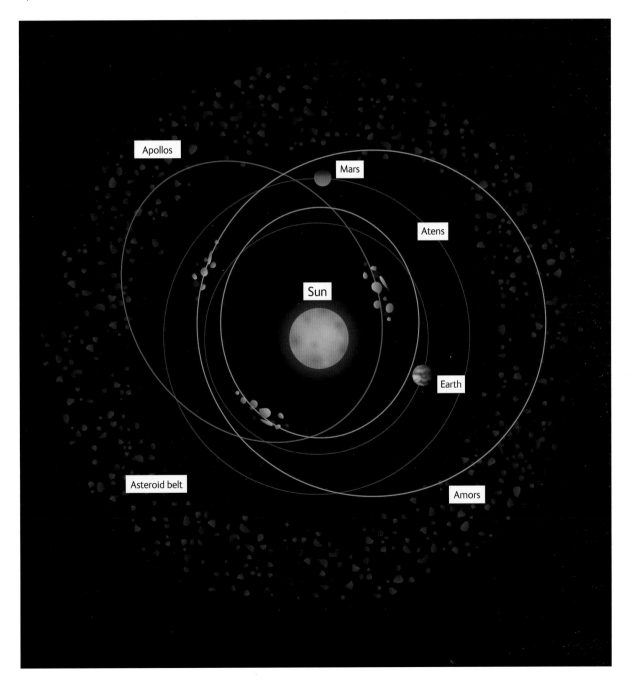

As of January 2018 a total of 17,508 near Earth asteroids is known, of which 1,886 are classified as potentially hazardous. For an object to be classified as potentially hazardous it must have an orbit that brings it within 750,000 km (466,028 miles) of the Earth's orbit at some point in the future. These are classified as hazardous based only on orbit. Further research must be carried out to determine the size of the body.

It is now thought that a global catastrophe (which would wipe out humanity) could be triggered by an impact of an asteroid or comet around 1 km (about ½ mile) across. The challenge is to map the orbits of all NEOs down to diameters of around 1 km (about ½ mile). This work is currently being undertaken by several international groups. One of the problems of tracking an NEO is that several observations over multiple orbits must be made before its final orbit is refined. Over the past few years there have been several reported 'near misses'. To reduce the danger of 'crying wolf' too frequently, the Torino Scale was adopted in 1999. This scale, analogous to the Richter Scale for earthquakes, is used for reporting the potential hazard of NEOs. By January 2018 there had been no known NEOs that had been rated above 0 on the Torino Scale.

BELOW Schematic representation of the Torino Scale established to report the potential hazard of an impacting NEO.

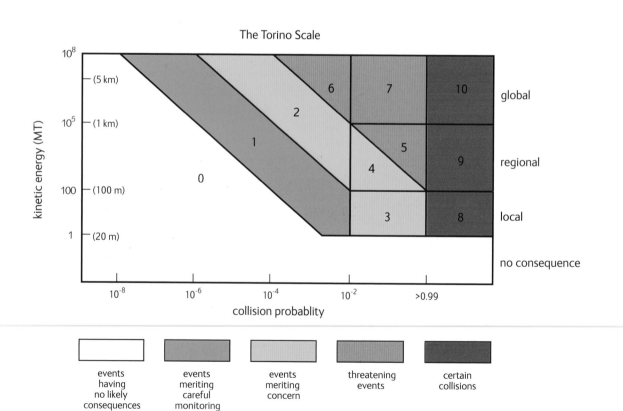

## THE CRETACEOUS–PALEOGENE BOUNDARY EVENT

Chicxulub Crater

Yucatan Peninsula

United States
of America

Mexico    Yucatan
Peninsula

Approximately 66 million years ago, at the end of the Cretaceous period, there was a dramatic drop in the numbers of species present on the Earth. This mass extinction has been linked with the collision of a huge meteorite with the Earth. The impact site was at Chicxulub in the Yucatan Peninsula in the Gulf of Mexico. The crater is now buried, but geophysical surveys estimate its diameter to be roughly 180–200 km (112–124 miles).

Environmental effects caused by an impact of these dimensions include massive fires ignited by heat radiated from the impact, and a darkening of the sky due to ejected rock dust and smoke followed by a rapid, global drop in temperature. In the case of Chicxulub, the impact was into sedimentary rock, including sulphate-bearing rocks. This would have resulted in tonnes of sulphur

ABOVE Radar topography of the 180–200 km (112–124 miles) Chicxulub crater on the Yucatan Peninsula. The dashed line shows the crater rim.

oxides being ejected into the atmosphere. The energy of the impact fused nitrogen and oxygen from the atmosphere into nitrogen oxides. As the temperature dropped, sulphur and nitrogen oxides washed out of the atmosphere as acid rain. These consequences affected the entire globe, not just the local region, and for an extended period of time. It is entirely possible that the global environmental changes caused the extinction of many species, including the dinosaurs, although this theory is by no means completely accepted by many palaeontologists.

# CHAPTER 4

# Sources and types of meteorites

ETEORITES ARRIVE ON EARTH from three well-determined sources. The vast majority (around 97%) of meteorites originate from asteroids; a few meteorites come from the moon (see p.86), and a few come from the planet Mars (see p.91). There is evidence to suggest that a very rare type of meteorite may come from comets, but this is still an area of scientific debate (see Chapter 5).

The most common class of stone meteorites is the chondrite; these samples are little changed since they formed, early in our solar system's history (see p.56). Other stone meteorites, the achondrites, show signs of melting and recrystallization and record igneous processes on their parent bodies (see p.78). Iron and stony-iron meteorites have also been melted and provide us with insights into the first stages of planet-forming processes (see p.75).

## ASTEROIDS AND METEORITES

Stony-iron meteorites, iron meteorites, chondrite meteorites and most of the achondrite meteorites originate from asteroids. Asteroids, or minor planets, are small bodies of rock and metal which are the remains of the first materials to form in our solar system, nearly 4.6 billion years ago. It is a common misconception that asteroids are an 'exploded planet'. In fact, asteroids are bodies that were never part of a single parent planet. The effect of Jupiter's gravity kept the asteroid bodies apart and prevented them from aggregating. Over half a million asteroids have already been observed, and the total number is likely to be much higher. However, if you were to add all the asteroids together, their total mass would be less than that of the moon.

Most asteroids orbit the sun in a belt between the orbits of Mars and Jupiter and they form a boundary between the inner, rocky planets (also known as the terrestrial planets), and the outer, gas and ice-rich planets. These asteroids are known as main belt asteroids. They have roughly circular orbits and have been in stable orbits within the asteroid belt since they formed. However, other asteroids have more elliptical orbits, some of which approach or even cross the orbits of the

OPPOSITE An optical microscope image of the Krymka chondrite showing numerous rounded objects, known as chondrules. The different colours in the image show the different mineral grains that make up the chondrules. The horizontal field of view is 3.7 mm ($^3/_{20}$ in).

RIGHT Photomontage of asteroids Ida (left), about 30 km (19 miles) long x 10 km (6 miles) wide, and Gaspra (right), about 17 km (10½ miles) long.

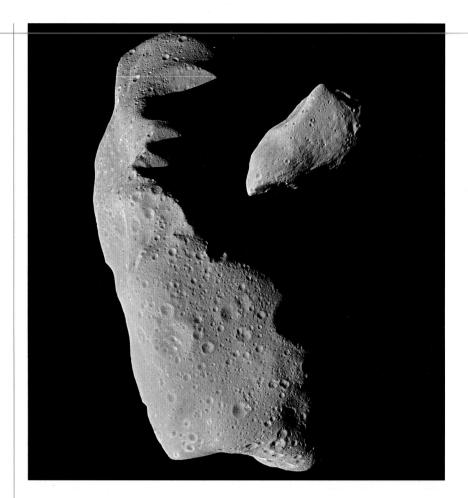

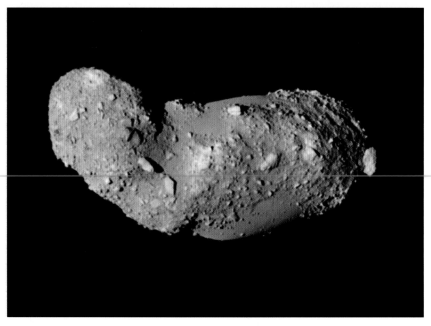

RIGHT Image of the asteroid Itokawa as imaged by the Hayabusa (Japanese) space mission. This is a very small asteroid of about 500 m (1640 ft) long. The surface of the asteroid shows a rubbly texture.

inner, rocky planets (Mercury, Venus, Mars and Earth) and occasionally they impact the surfaces of these planets. Asteroids with orbits that cross that of the Earth are known as near earth objects (NEOs, see p.46).

Geological history records past collisions between asteroids and the rocky planets, and impact craters scar their surfaces. On Earth large impacts are often associated with significant changes in the planet's biodiversity, through extinctions and evolutionary change (see p.48). Space missions that have visited and imaged asteroids, such as the NASA NEAR (Near Earth Asteroid Rendezvous) mission and the Japanese Hayabusa mission, show that asteroids have irregular shapes and

BELOW Distribution of asteroids in the inner solar system.

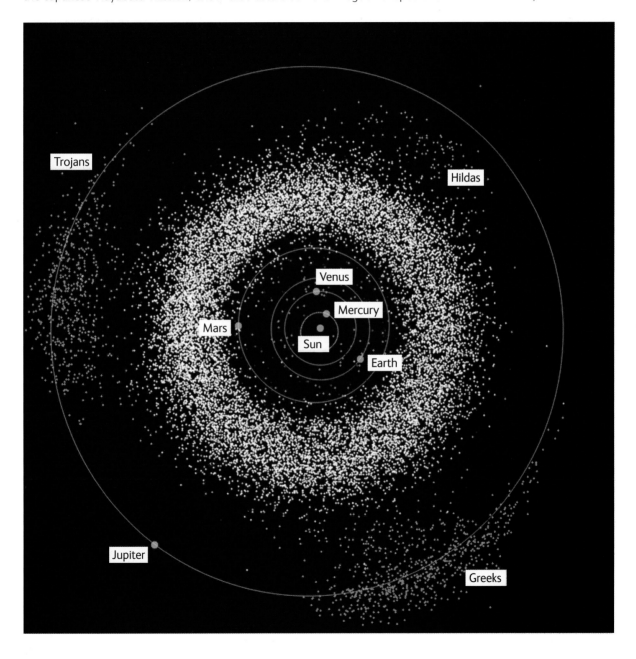

their surfaces are highly cratered. This suggests that asteroids have suffered many collisions and impacts during their long lives. These collisions provide the majority of the meteorite material we have on Earth today (see Chapter 3).

As we have learned in Chapter 2, calculations of the orbits of fireballs that subsequently produced recovered meteorites provided definitive proof that these meteorites originated in the asteroid belt. Observations of asteroids by telescopes have also provided evidence of the intimate link between asteroids and meteorites.

As is the case with all other solar system bodies, asteroids reflect sunlight from their surfaces. The colours, or spectra, of this reflected light are dependent on the composition of the asteroid. Spectroscopic observations of asteroids by telescopes on Earth, and from spacecraft that have visited asteroids, show that they can be subdivided into different types on the basis of their spectral signatures. Their surfaces can be dominated by rocky, metallic, or rock and metal, and many are thought to contain carbonaceous material. From these basic characteristic spectroscopic signatures, there appear to be at least 14 different asteroid groups.

Asteroids can also be divided into families based on their orbital characteristics. Asteroids in the same families tend to share the same compositions and so they are assumed to have formed through the break-up of a larger, pre-existing asteroid. Scientists can also compare spectroscopic data from asteroids with those of meteorites

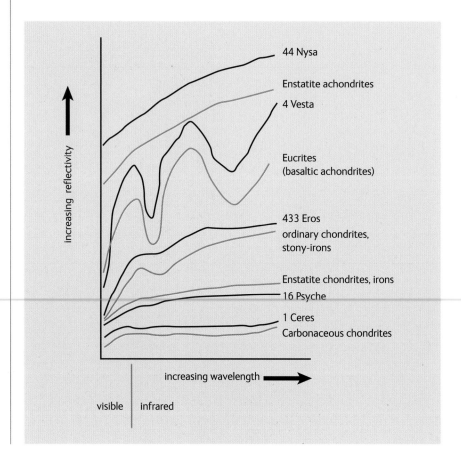

RIGHT Spectra of asteroids show some similarities with the measured spectra of meteorites. The strong similarity between the basaltic achondrites (eucrites and also the related howardites and diogenites) suggest that these meteorites were derived from 4 Vesta.

## HOW ARE ASTEROIDS NAMED?

The first – and largest – asteroids to be discovered were given names from classical mythology, such as Vesta, Ceres and Juno. The naming of meteorites is now governed by the International Astronomical Union (IAU). When an asteroid is discovered it is given a provisional name. The letter after the year designates the half-month period within the year it was found. 'A' would be for the first half of January, 'B' for the second half, and so on. For instance, the 'E' in 1981 ES4 indicates that it was discovered in the first half of March. The second letter plus the number denote the order of discovery during that half-month period. So, for our example above, it was the 118th asteroid observed in that 2-week period because 4 x 25 (number of letters in the alphabet minus 'I') + 18 ('S') = 118.

An asteroid is only formally named once its orbit has been confirmed; it is then given a number, according to the date when the data on its orbit were determined. Finally, a name can be assigned. For example Vesta is the fourth asteroid discovered, while (9007) James Bond is the 9,007th asteroid discovered. Although there are over 200,000 asteroids with numbers, fewer than 20,000 have been given names. The choice of name is up to the discoverer(s); however, the convention is that NEOs are still given traditional names whereas main belt asteroids are often named after individuals, some of whom may (or may not) be famous. The name must be inoffensive and not related to military activities. Once a name has been chosen by the discoverer(s) the IAU votes to accept and publish the names. Asteroids have been named after musicians, such as (1034) Mozartia and (1815) Beethoven, as well as (4147) Lennon, (4148) McCartney, (4149) Harrison and (4150) Starr; artists including (6676) Monet and (4511) Rembrandt; writers such as (2985) Shakespeare and (3047) Goethe; scientists such as (697) Galilea, (662) Newtonia, (8000) Isaac Newton, (1991) Darwin and (2001) Einstein; plus other miscellaneous people such as (5497) Sararussell and (7635) Carolinesmith.

on the Earth. There are good matches between some meteorite classes and some asteroid types. The howardite, eucrite and diogenite (HED) group of meteorites (see p.69) have similarities in their chemistries, suggesting that they formed in the same asteroid. Spectral studies of the HEDs show a very strong match with one of the largest asteroids known, 4 Vesta. It seems very likely that these meteorites are derived either from this asteroid or perhaps from the 'Vestoids' – asteroids that are believed to have originally been part of 4 Vesta, and which were ejected from its south pole region in a very large impact. This relationship was confirmed by the NASA Dawn mission, which analyzed the composition of the surface of Vesta in 2011–2012.

The JAXA Hayabusa mission visited the asteroid Itokawa (see p.52), a common 'S' type, and returned some fragments to Earth. The material returned from this mission proved that this asteroid type is the parent for the most common meteorites, the ordinary chondrites. However, there are also inconsistencies between the distribution of asteroid types and the classes of meteorites. For example, the most abundant asteroid type (spectral class C) has very similar spectral characteristics to the stony, carbonaceous chondrite meteorites. However, the carbonaceous chondrites are not particularly abundant on Earth, representing only around 3% of known meteorites. The inconsistencies in the spectra of known meteorite classes, with the spectra of asteroids, suggest that meteorites originate from asteroids that are not typical of the total asteroid population. To put it another way, the meteorites we have on Earth are not representative samples of the asteroids. It is estimated that all the meteorites on Earth originated from around 100 parent asteroids. This represents just a tiny fraction of the asteroid belt.

# UNMELTED METEORITES AND PLANET FORMATION

Meteorites are divided into three main types: stones, irons and stony-irons, as discussed on p.5. Stony meteorites are further subdivided into the unmelted chondrites and the melted achondrites.

## CHONDRITES

Chondrites are thought to have formed at the same time and from the same material as the inner rocky planets of our solar system. Studying them can tell us about how the planets were made and about the materials that originally formed our own planet. They are a jumbled mixture of materials that date from the time the solar system formed. Most of the components are usually a few millimetres across, or smaller. The major constituents are chondrules – rounded, millimetre-sized, once-molten objects that formed during a fast-heating event. (Chondrules are named from the Greek word 'chondros', meaning 'seed' or 'little grain'.) Chondrites also contain shiny flecks of iron–nickel metal, plus occasionally pale-coloured patches of calcium and aluminium-rich inclusions (CAIs). The fine-grained matrix surrounding these objects is made up of a mixture of very fine-grained crystalline material along with amorphous glassy material. The matrix contains tiny fragments of chondrules, along with organic material and ancient pre-solar grains (see p.61).

Chondrites are divided into four main classes: ordinary (O), carbonaceous (C), enstatite (E) and rumuruti (R) chondrites. These four may have formed at different distances from the sun. Ordinary chondrites, as the name implies, are the most common type of meteorite. They account for 86% of known meteorites. Carbonaceous chondrites are much less common, but can perhaps tell us even more about the origins of the solar system. They have a composition that is richer in volatile substances (such as carbon-bearing compounds) than the ordinary chondrites. In fact, the composition of some of them is extremely similar to that of the sun, minus the gaseous hydrogen and helium that make up the vast majority of solar material. For this reason they are thought to represent the composition of the inner planets within the solar system, and their composition is used by geochemists to compare to terrestrial rocks. Enstatite chondrites are notable for having very oxygen-poor compositions, so much so that elements that usually bond with oxygen in other meteorites (and on Earth) have formed compounds with sulphur. Rumuruti chondrites, in contrast, have more oxygen for bonding than the other meteorite groups, and are thus devoid of metal. In addition to the four main types of chondrite, there are also several chondrites that do not fit very well into any of the known groups, such as the small Kakangari chondrite grouplet.

As well as differing in chemistry, the different chondrite groups can be distinguished by their oxygen isotopes. Oxygen has three isotopes: oxygen-16 (the most abundant), oxygen-17 and oxygen-18. On formation, the major meteorite groups contain different proportions of these isotopes. The proportion then sometimes changes, depending on different factors, such as how much alteration by heat or water the meteorite has suffered. These variations are an important tool used in meteorite classification.

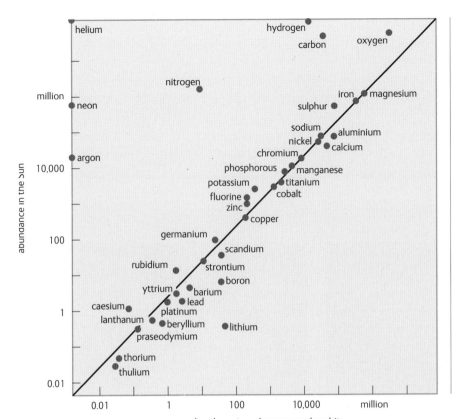

Except for gases such as hydrogen and helium, the chemical composition of carbonaceous chondrites such as Ivuna and the Sun is very similar.

BELOW LEFT The Parnallee ordinary chondrite, containing many chondrules.

BELOW RIGHT The Ivuna carbonaceous chondrite. This type of meteorite is the richest in volatile elements.

## PLANET FORMATION

The mechanism by which planets formed can be investigated in four ways: by looking at rocks that formed very early in our solar system's history, such as chondrites; by observing young stars to look for signs of planet formation; by space missions to explore our solar system; and by theoretical models of planet formation. All approaches have contributed to what we know of our planet's origins. To understand what meteorites can tell us about the formation of the planets, we must first look to theoretical and observational information about planet formation.

## FORMATION OF THE SOLAR SYSTEM

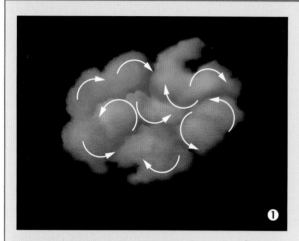

**❶**

A turbulent cloud of interstellar dust and gas collapses.

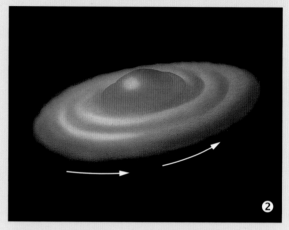

**❷**

The dust and gas cloud form a spinning disc.

**❸**

The temperature and pressure of the disc increase towards the middle. Eventually, the temperature in the centre is hot enough for fusion of hydrogen into helium to begin. The central star is born.

**❹**

The remaining dust and gas clump together, gradually sweeping up all the debris into planets.

**A HISTORY OF PLANET FORMATION THEORIES** Since Copernicus' discovery that the planets orbit the sun, there have been three major theories about how the planets formed. One of the most popular theories, first postulated by the astronomer James Jeans (1877–1946), was the ejection hypothesis. This states that the material from which planets were formed was ejected from the sun, as a filament of gas, following a collision with a giant comet or another star. The theory was later rejected because planets are relatively enriched in the hydrogen isotope deuterium, and the element lithium, which would be destroyed in a star. A second major theory, developed in the 1930s, was the capture hypothesis, which states that passing material was captured by the early sun's gravitational pull. Although the capture hypothesis has never been disproved, recent astronomical observations make the third, and oldest theory – the nebula hypothesis – more attractive. By the nebula theory, the solar system formed from a rotating, flattened disc. This idea was first proposed by the Prussian philosopher and astronomer Immanuel Kant in 1755, and further developed by Pierre-Simon, Marquis de Laplace, in 1796. It remains the most widely accepted theory of planet formation to this day, because it is compatible with both the observational astronomical evidence and evidence from meteorites.

The modern-day nebula theory pictures the sun and planets forming at the same time, from the same dense cloud of interstellar dust and gas. As most of the cloud, collapsed by gravity, swirled inwards to form the sun, the remains of the infalling cloud began to form a rotating disc, called an accretion disc or solar nebula, around the central star, from which the planets were born.

BELOW Young stars imaged using new-generation telescopes provide clues about the origins of our own sun and solar system, and show how the planets may have formed from a dusty disc surrounding the young sun.

**OBSERVATIONAL EVIDENCE: PROTOPLANETARY AND DEBRIS DISCS** Most very young stars, less than about 5 million years old, have discs of dust and gas, called protoplanetary discs. These discs become less and less common around progressively older stars, as they dissipate or form planets. The dust is mainly a mix of silicates, carbonaceous material and ices, much like our own solar system at the time of its formation, and the gas is mostly hydrogen and helium.

Around 15% of nearby older stars also have discs around them. These discs have proportionally large amounts of dust and small amounts of gas. One of the best-known examples is Beta Pictoris. Since dust is expected to fall into the star over a short timescale, these dusty discs around old stars are probably continually replenished with new dust from asteroids and comets. Some of the discs are ring-shaped, with a hole close to the star. What is in the hole? Paradoxically, telescopes enable us to detect large clouds of the tiny dust grains, even though at the same range individual planets, asteroids and comets remain invisible to us. The inner holes could be full of planets. In fact, our own solar system, if viewed from a neighbouring star, would also look like a star surrounded by empty space in place of the planets, but clearly showing the dust ring beyond Neptune (this dust ring is known as the Kuiper Belt and is described in more detail in Chapter 5).

Recent advances in telescopes are finally allowing us to see planets around other stars. It seems our suspicions that many dusty discs go on to form planetary systems are correct. Many mature stars appear to have one or more planets orbiting around them. A beautiful example of a young star system is HL Tauri (opposite). Here, the dusty disc contains gaps, where we presume new planets are forming and sweeping up grains in their way.

**FORMATION OF THE TERRESTRIAL PLANETS: THE STORY FROM METEORITES** Much of our knowledge about the formation of the four inner, rocky planets (the 'terrestrial planets') comes from meteorites. Meteorites reveal a complex picture of planet growth. The first stage is the coagulation of dust grains, which are initially only 1 millionth of a metre (3.9 hundred thousandth of an inch) across or less. Dust coagulation may have been aided by heating events that allowed the dust to melt and stick together to make chondrules. These objects then collided and – perhaps because they were moving at such gentle relative speeds that they could accrete from static forces, or using organic material as glue – formed boulders of 1 m to 1 km (3¼ ft to about ½ mile), called planetesimals. Once the planetesimals reached a certain size, they began to have a significant gravitational field. This made their growth much easier and quicker, as other planetesimals were attracted to them and their growth progressed by impacts of these relatively large bodies. This stage is called 'runaway growth'. Although this broad outline of how planets formed seems to fit most of the features seen in meteorites, deciphering the history recorded in meteorites to understand the details of planet formation proves much more difficult. Some important

OPPOSITE HL Tauri, a protoplanetary disc. In the centre of the image is a young star surrounded by rings of dust, showing as orange. In the gaps between the dust rings, planets may be forming.

Barred olivine chondrule in the Bishunpur chondrite. These are thought to have formed at temperatures up to 2,100°C (3,812°F). This chondrule is about 0.5 mm ($^1/_{50}$ in) in diameter.

questions remain. How exactly did chondrules form? How long did the process of planet formation take? Did the planets always exist in the orbit they are in now, or did they move inwards or outwards? Ongoing studies of meteorites are helping to address these questions.

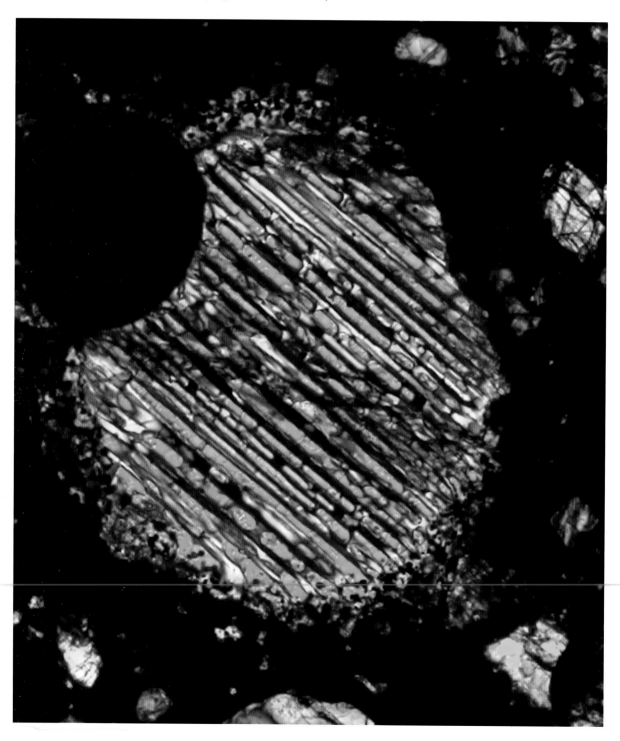

ABOVE LEFT Radial pyroxene chondrule in the Bishunpur chondrite. Similar to barred olivine chondrules, radial pyroxene chondrules are thought to have formed by the complete (or nearly complete) melting of precursor materials, at high temperatures. This chondrule is approximately 0.5 mm ($1/50$ in) in diameter.

LEFT Porphyritic olivine and pyroxene chondrule in the Chainpur chondrite. Porphyritic chondrules are the most abundant types of chondrule observed in ordinary chondrites. This chondrule is approximately 1.5 mm ($3/20$ in) in diameter.

## COMPONENTS OF CHONDRITES

**CHONDRULES** These are near-spherical, typically millimetre-sized stony objects, which probably formed by a sudden, flash heating event at temperatures of over 1,400°C (2,552°F) followed by fast cooling, and solidification of the resulting molten droplets. Chondrules are mostly made up of the magnesium, iron, silicon and oxygen-rich minerals olivine and pyroxene. The exact details of the process and the nature of the heat source that produced these objects early in solar system history are some of the long-standing mysteries in meteoritics. It is important to find the answers to these questions, since this may tell us how the very first stage of planet formation, the accretion of tiny dust particles into larger millimetre to centimetre-sized balls, took place. By looking at the chemistry and texture of chondrules we can place a few constraints on how they formed. For example, many chondrules appear to contain 'chondrules within chondrules', suggesting that they experienced a high-temperature heating event more than once. The main theories for chondrule formation involve a shock wave in the early solar system or collision between planetesimals, and these theories are discussed further below.

BELOW These images show changes over only a five-year period in the disc and jets of this newborn star, which is about half a million years old. The young stars are obscured by a dark disc of dust. Because of changes in the local magnetic field, violent jets of material are ejected from near to the central star.

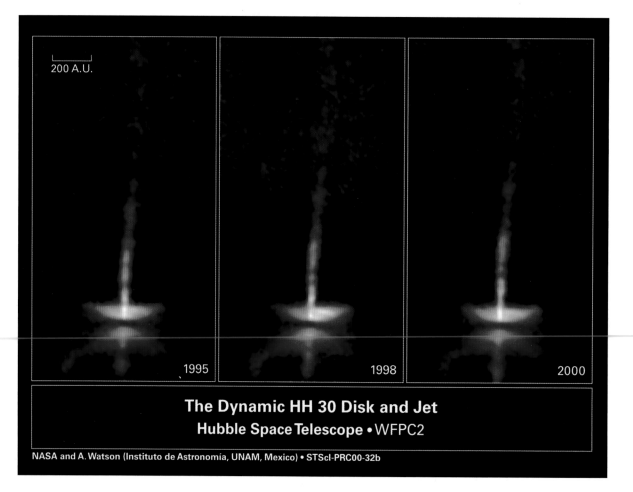

200 A.U.

1995

1998

2000

**The Dynamic HH 30 Disk and Jet**
**Hubble Space Telescope • WFPC2**

NASA and A. Watson (Instituto de Astronomía, UNAM, Mexico) • STScI-PRC00-32b

LEFT The Barwell meteorite, showing a rock fragment formed by melting of a pre-existing planetary body; this indicates that planetesimals were present when chondrules formed, strengthening the collision theory.

The shock wave theory argues that in the early solar system there would have been waves of high pressure passing through the inner region where the planets and asteroids formed. The shock of being hit by this wave of pressure would be enough to heat and melt dustballs and cause them to form chondrules. One of the main lines of evidence for the second theory, planetesimal collisions, is that some chondrites contain rare rock fragments that have undergone melting themselves. It is argued that these rock fragments came from planets that must have existed before the chondrites formed. Chondrules may then have formed as a product of high-energy collisions between planetesimals just tens or hundreds of kilometres in diameter. In truth, both (and maybe more) processes are likely to have formed chondrules, and the challenge for us now is to understand which were most important.

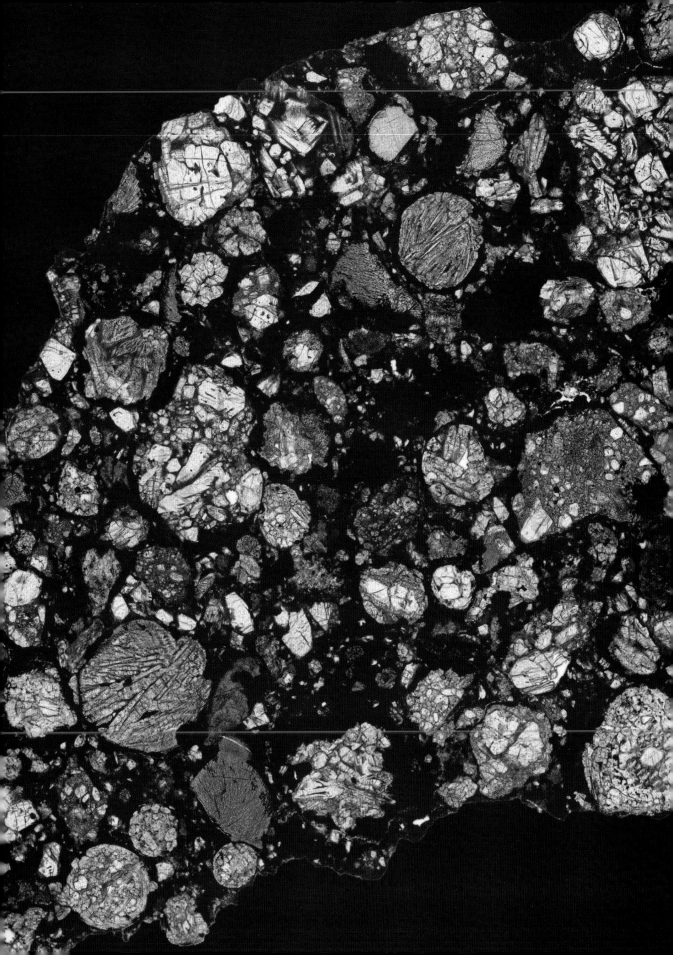

LEFT The Beddgelert (H5) ordinary chondrite clearly showing flecks of iron-nickel metal.

**METAL** One thing that best distinguishes chondrites from terrestrial rock is that they almost always contain flecks of metal – a mixture of iron and nickel. This makes chondrites slightly attracted to a magnet, and provides a good test for meteorites. The presence of metal within the stony matrix is extremely important. It implies that the rock has not been melted since the earliest times in the solar system, because the molten metal has not sunk to the core of its parent asteroid, as has happened on Earth. The quantity of metal in chondrites is quite variable, from 0% to 20%, and metal content provides an important means of classifying chondrites of different types.

**CAIS** In February 1969 a large shower of stony meteorites fell near the village of Pueblito de Allende, Mexico. The meteorite turned out to be of the rare carbonaceous chondrite variety. Many of these stones were quickly retrieved and over 2 t of material was collected and eagerly studied by meteorite scientists all over the world. One immediately obvious and unusual feature of the Allende meteorite is that on each surface millimetre- to centimetre-sized, irregularly shaped, pale-coloured inclusions are clearly visible – these are CAIs.

CAIs have a composition that is subtly different to chondrules. They are typically much more enriched in the high-temperature elements calcium and aluminium, and contain less abundant silicon, magnesium, iron and more volatile substances. The ages of several CAIs have been measured and they have been found to be the oldest known solids in the solar system, dating from around 4,568 million years ago. They are quite common in

OPPOSITE An optical microscope image of the Chainpur meteorite showing the very clear chondritic texture. Different chondrule types can be seen in this image, including barred olivine, radial pyroxene and porphyritic chondrules. This image is 5.3 mm ($^{1}/_{5}$ in )across so the largest chondrules are about 1 mm ($^{1}/_{25}$ in) in diameter.

BELOW The Allende carbonaceous chondrite, partly covered in jet-black fusion crust. This contains numerous white inclusions called CAIs. This stone is about 10 cm (4 in) across.

carbonaceous chondrites, but extremely rare in ordinary and enstatite chondrites. They are also different from chondrules in their oxygen isotope composition. The formation mechanism of CAIs is unknown, but it may be similar to that of chondrules, although many CAIs seem to have been heated for longer, and cooled more slowly. An important difference between the two types of objects is that not all CAIs formed from molten droplets. A few CAIs have fragile, intricate textures like snowflakes, suggesting that they condensed directly from a vapour to a solid without ever being molten.

RIGHT A CAI from the Leoville carbonaceous chondrite. It is 2 cm (¾ in) across.

BELOW RIGHT A cluster of millions of diamonds, separated from the Allende meteorite. Some scientists believe these are pre-solar.

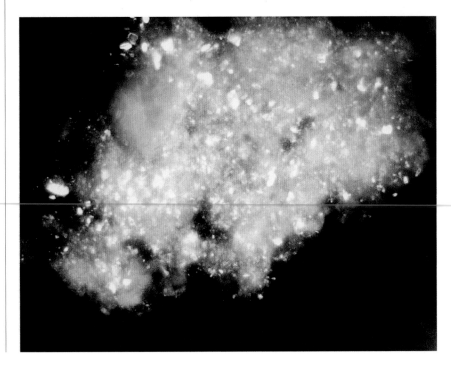

**PRE-SOLAR GRAINS** The matrix of chondritic meteorites contains some tiny crystals of very resilient minerals, such as diamond, silicon carbide, corundum, silicon nitride and graphite. The isotopic composition of these grains is completely different to the composition of everything else in their host meteorites, or indeed anything on Earth or in the rest of the solar system. These 'pre-solar' grains are so named because they are thought to have existed before the solar system formed. They are ancient minerals that originally formed around stars such as red giants or supernovae. Stars such as these were our ancestors, providing the elements from which the solar system formed. The grains produced around these stars floated in interstellar space before being swept up and incorporated into our newly forming solar system. Pre-solar grains provide unparalleled opportunities to perform astrophysics in the laboratory, to learn about element- and grain-forming processes occurring in and around other stars. Studies of pre-solar grains have told us that the solar system is made up of some material from a supernova as well as material from several red giant stars. Dozens of stars separated by billions of kilometres contributed to our solar system, ultimately sitting side by side in a single meteoritic rock.

BELOW A pre-solar grain of silicon carbide, about 1 µm across, from the Murchison meteorite.

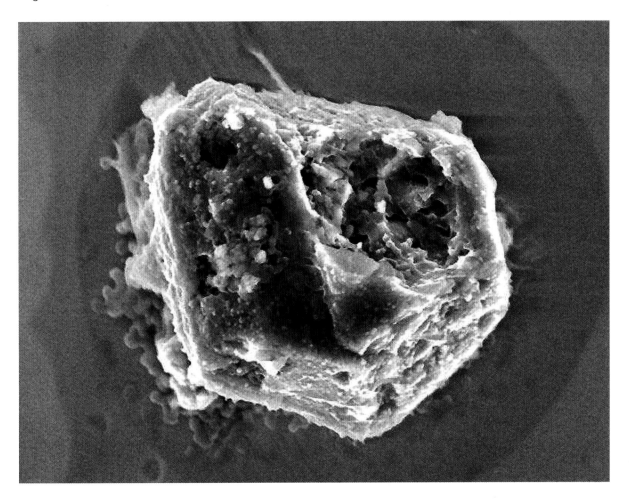

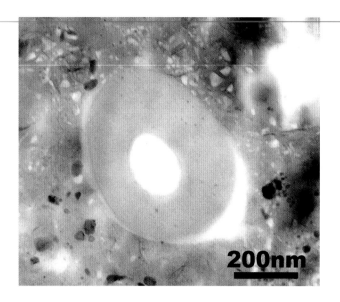

**200nm**

ABOVE Tranmission electron
microscope (TEM) image of carbon-
rich, organic globules in the Tagish
Lake meteorite. The composition
of the deuterium and hydrogen in
these objects indicate that they
have a pre-solar origin.

**ORGANIC COMPOUNDS** Some meteorites contain organic compounds made up of the elements carbon, hydrogen, oxygen and nitrogen. These compounds, found most abundantly in carbonaceous chondrites, are the main constituents of living things. Those found in meteorites are thought to have formed abiotically, that is, no life was involved in their formation. Some formed in interstellar space prior to the formation of the solar system, some within the solar nebula and perhaps some in the asteroid parent. Meteorites containing organic material probably fell to Earth before life flourished here, and may indeed have been an important first step to life beginning on Earth, by providing the basic vital ingredients needed to create living things.

## HEATING OF YOUNG PLANETESIMALS

Nearly all chondrites, and certainly all melted meteorites (see p.73), show evidence of having been heated since they formed into a planetesimal. In the case of achondrites and iron meteorites, this heating event was clearly very severe, and enough to cause the rock to melt. In the case of chondrites, the extent of their alteration since their formation is highly variable and this provides a secondary classification scheme for these meteorites.

Chondrites are classified by number, from 1 (unheated but water damaged) to 6 (very severely heated, up to 950°C or 1,742°F). A few rare meteorites seem to have experienced very little heating since they formed and apparently remain almost unchanged. The most 'pristine' meteorites are usually those of Type 3. These have experienced minimal heating and little alteration by water. Others (Types 1 and 2) contained a lot of water-ice when they formed. Heating on these bodies had the effect of melting the water, leaving the rocks water-damaged but not extensively heated.

However, the majority of meteorites (of Types 4 and higher) have textures that suggest they have suffered changes brought about by heat. Heating has altered the mineralogy and textures of the rocks and has caused the constituent elements to move around and become well mixed. A tiny proportion of the heat in planetesimals is simply gravitational, as the components of the chondrites collapsed to form a larger body they released heat. However, this process cannot account for much of the heat required to produce the effects that are observed. We must look to a different mechanism, and the most likely heat source appears to be from the decay of radioactive isotopes with extremely short half-lives (i.e. unstable isotopes that decay quickly, relative to geological processes, producing energy). Analysis of meteorites shows that they once contained such radionuclides.

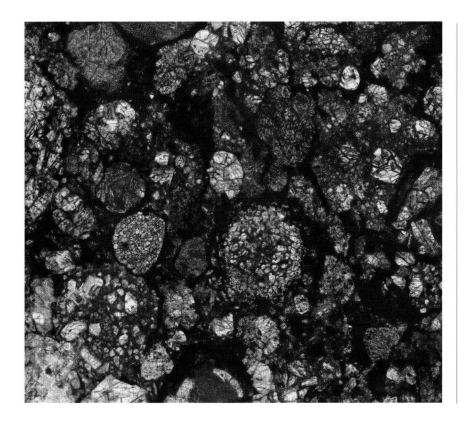

LEFT An optical microscope image of the Parnallee (Type 3) chondrite that has experienced little heating. The chondrules are clear and well-defined. The field of view is 3 mm ($^1/_{10}$ in).

BELOW An optical microscope image of the Barwell (Type 6) chondrite. This meteorite has experienced a significant amount of heating, and the chondrules are not so well defined as for the Parnallee meteorite (see above). The field of view is 5 mm ($^1/_5$ in).

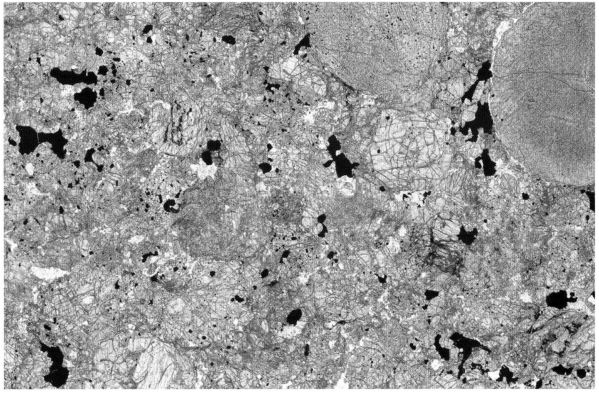

## HOW OLD ARE THE METEORITES, PLANETS AND THE SOLAR SYSTEM?

Ages are measured using radioactive 'clocks'. By 'age', in the case of melted meteorites, we mean the time since the minerals cooled from a molten state to their current solid state. In the case of unmelted chondrites, this can either mean the age of the individual constituents of the meteorite or the age at which the constituents coagulated together.

Elements sometimes have more than one isotope, particles of the same element that have different masses. While many isotopes are stable and can survive intact indefinitely, some are radioactive and break down to form different isotopes and elements. The half-life – which measures the rate at which these radioactive isotopes disintegrate – can be measured, and so by measuring the amounts of both the parent isotope and the decay products of its daughter isotope present in a rock, the age of the rock can be determined. The age determined is the time from when the rock was last hot enough for its constituent isotopes to be able to move around freely. The isotopes that are most commonly used for age dating of ancient solar system rocks are rubidium-87 (which decays to strontium-87 with a half-life of 49 billion years), samarium-147 (which decays to neodynium-143 with a half-life of 100 billion years) and two uranium isotopes that both decay to lead isotopes. Since most meteoritic fragments of asteroids yield similar ages (around 4.57 billion years), older than any rocks that originate on Earth, this is thought to be the age of the solar system and the planets in it.

Isotopic studies of meteorites show that they contained short-lived isotopes when they formed. The most important heat source may have been the isotope aluminium-26. We know that this isotope was once active inside meteoritic material because today the aluminium-rich minerals in these meteorites contain excessive amounts of magnesium-26, the decay product of aluminium-26. Most radioactive isotopes were present in the solar system at the levels expected for interstellar material, but some isotopes, such as $^{10}$Be, have much higher abundances than expected. A proportion of these short-lived isotopes may have formed in a stellar event, such as a supernova, that must have happened very shortly before the formation of the solar system or close to the young sun and were transported to the asteroid and planet-forming region in jets. However they formed, they are important because they can be used as extremely sensitive clocks of early solar system processes.

Meteorites (and their components) that have a large amount of decay products must have contained a large quantity of radioactive nuclides when they originally formed. The more radioactive isotopes that rocks initially contained, the older they are likely to be – in younger rocks, these isotopes would have mainly decayed away by their formation time. For example, the work of scientists Alex Halliday and D.-C. Lee on the short-lived isotope hafnium-187 has been used to infer that the planet Mars formed relatively quickly, within a few tens of millions of years of the start of the solar system. This is because isotopic studies show that Martian rocks contained this isotope when the planet formed. In contrast, the Earth formed much later, when all the hafnium-187 had already decayed away. These, and other isotopes such as 26Al, can therefore be used as a chronometer of important events in the early solar system. They can provide information, for example, on how long the protoplanetary disk lasted and how long planets took to form.

# MELTED METEORITES

In contrast to meteorites that have experienced only limited heating or some aqueous alteration, like those previously discussed, some meteorites come from asteroids that have totally or partially melted. There are three types of melted meteorites: irons, stony-irons and achondrites. These meteorites come from planets or asteroids that have undergone differentiation. In this section we discuss those that come from asteroids.

When a chondritic asteroid is heated to its melting point (between 1,000 and 1,300°C or 1,832 and 2,372°F depending on its composition), the distribution of minerals changes dramatically. The silicate material will melt and the flecks of metal will coalesce; and, because the metal is heavier than the silicate material, it will sink to the centre or core of the body even under the low gravitational forces of an asteroid. As the body gets hotter, the silicate minerals will start to melt. These form basalt, a rock composed of plagioclase and pyroxene, and this melt will rise to form the crust of an asteroid. Minerals with a medium density will form a lower layer called the mantle. This process of metal and silicate separation is called differentiation.

## DIFFERENTIATION PROCESS

Differentiation is the name given to a process whereby a planet or asteroid transforms from a homogeneous starting material into a core, mantle and crust based on density variations of minerals in the starting materials.

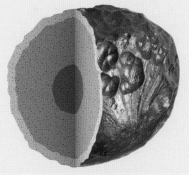

A primitive, chondritic body before heating begins.

A primitive body when heating and melting has begun; the heavier minerals (in red) start to sink to the middle (core) of the body, while lighter minerals (pink) begin to float to the surface. The medium weight minerals (brown) have been heated and melted enough to allow the others to sink or float.

The body has completely separated into a core (red), mantle (brown) and crust (pink). This process, called differentiation, has taken place on all the planets in the inner solar system, including Earth's moon. It has also occurred on a number of asteroids.

ABOVE A piece of the Henbury iron meteorite, 28 cm (11 in) across, which once formed part of an asteroid's core.

Based on meteorite studies, we know that some asteroids have undergone differentiation, but there are still many questions remaining about the process. For instance, how big does an asteroid need to be to melt? Was differentiation of asteroids a common process in the early solar system? If it was very common, did planets form from differentiated planetesimals, rather than primitive, unmelted ones?

The iron meteorites are composed of iron with varying amounts of nickel and are thought to represent material from the cores of differentiated asteroids. There are two types of stony-iron meteorite. The first, called pallasites, are composed of a green mineral (usually olivine) embedded in a metallic matrix. These types of meteorites are thought to be from the core–mantle boundary of differentiated asteroids, although an alternative theory suggests they were born in an impact. The other type of stony-iron meteorite is called a mesosiderite (which means 'half iron' in Greek). These meteorites are characterized by a jumbled mixture of basaltic and metallic material and clearly formed when two molten asteroids collided. The stony melted meteorites, the achondrites, represent material from the crusts and mantles of melted asteroids and planets. These meteorites are called achondrites to distinguish them from the other type of 'stony' meteorites, the chondrites.

# IRON METEORITES

Most iron meteorites come from the cores of differentiated asteroids and are sometimes called magmatic irons. Some iron meteorites, termed non-magmatic, come from asteroids where the process of differentiation was interrupted. All iron meteorites are classified according to chemical composition and structural features. Iron meteorites are composed primarily of iron, with varying amounts of nickel, sulphur, carbon and phosphorus. On the basis of the abundance of nickel and elements that occur in trace amounts (at levels of parts per million or less), around 85% of iron meteorites fit into one of 13 groups (a group contains five members or more), and probably come from 13 different parent bodies. The remaining 15% cannot be grouped so readily and may originate from up to 50 different parent bodies.

The structure of iron meteorites is seen primarily in the Widmanstätten pattern, a texture that forms during cooling from 700°C to 450°C (1,292°F to 842°F) and, interestingly, only when the iron has become solid. When elements move between minerals while a rock is solid, it is called diffusion. To form the Widmanstätten pattern, the diffusion had to occur very slowly. This would happen if the asteroid core was well insulated and took a long time to cool. The Widmanstätten pattern is only revealed when the sample is cut and etched with a mixture of dilute nitric acid and alcohol. The pattern is named after Count Alois von Widmanstätten, who was credited with its discovery in 1808. However, the pattern was also discovered independently, some years earlier, by the chemists Edward Howard (see p.16) and William Thomson.

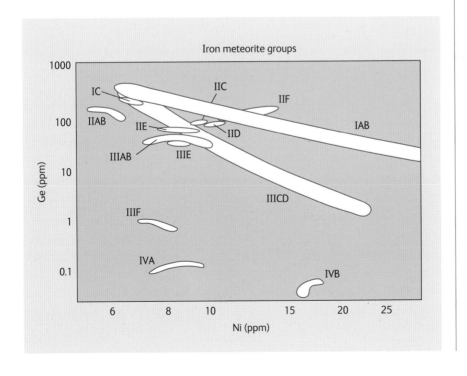

LEFT Simplistic germanium (Ge) vs nickel (Ni) plot in different iron meteorite groups.

RIGHT Iron meteorites, when sliced open and etched with acid, typically show a distinctive criss-cross pattern called a Widmanstätten pattern. This slice of Canyon Diablo is 15 cm (6 in) across.

The iron and nickel in these meteorites form two different minerals, one that is poor in nickel called kamacite and the other rich in nickel called taenite. Kamacite typically contains less than 10% nickel, while taenite can have an abundance of nickel that ranges between 25% and 50%. At high temperatures, when the metal is molten, the iron and nickel can easily mix, but when the metal becomes solid (or crystallizes) and cools these elements will form the characteristic criss-cross pattern. The thickness of the pattern gives information about the cooling history of these meteorites. A thicker pattern will form if the meteorite was cooling very slowly, which might be the case if it were in the middle of the core or under a thick silicate mantle. A thinner pattern forms if the cooling rate is more rapid. Iron meteorites also have other minerals besides the kamacite and taenite. The most common is troilite, a sulphide of iron that is often found as rounded nodules distributed randomly throughout many iron meteorites. Minerals containing carbon and phosphorus are also quite common.

## STONY-IRON METEORITES

Stony-iron meteorites are so named because they are a mixture of stony and iron material. As discussed above, there are two types of stony-iron meteorites. One group is called the pallasites. The name pallasite comes from the German scientist, Peter Simon Pallas, who described the Krasnojarsk pallasite in 1772, before it was known that this sample came from outer space. Pallasites, when sliced and polished, are perhaps the most beautiful of all the meteorites because of the mix of metal and centimetre-sized, gem-quality olivine that is called peridot. These meteorites are thought to have originated at the core–mantle boundary of large asteroids. The chemical composition of the metal in a few of the pallasites is similar to some iron meteorites, which suggests that they might have originated on the same type of asteroid.

LEFT The Esquel pallasite, composed of gem-quality olivine crystals embedded in metal.

The other stony-iron group is called the mesosiderites. They are composed of half iron and half silicate. In contrast to the pallasites, the silicate material found in mesosiderites is composed of igneous (or previously molten) rock and mineral fragments that formed on the crusts of differentiated asteroids. The way the metal and silicate are mixed up together gives a clue to how they formed. Because of the angular nature of the silicate fragments, the most likely formation scenario for mesosiderites is collision between two differentiated, and possibly molten, asteroids. The oxygen isotope composition of many of the rocky fragments in mesosiderites is identical to those found in the achondrite HED meteorites (see p.81), suggesting a link between these two very different meteorite types.

## MELTED STONY METEORITES OR ACHONDRITES

There are three groups of stony meteorites that are not classified as chondritic, because of their lack of chondrules and metal. Their textures are igneous (compare to chondrite textures in the figures on p.71), which means they crystallized from a molten state. The three groups are not related to each other and thus represent the crusts or upper mantles of at least three different types of asteroids.

**ANGRITES** Angrites is a small group of meteorites (only 28 members by December 2017) that are composed primarily of pyroxene and plagioclase, the mineralogy of a rock type called basalt. Basalt is a very common volcanic rock on Earth and is found in association with island volcanoes such as those in Hawaii, Iceland and the Aeolian islands off the Mediterranean coast of Italy. The angrites is an enigmatic group. The coarse textures of these meteorites indicate they may have cooled relatively slowly. The zoning patterns of elements in the minerals, however, indicate relatively rapid crystallization. The debate is ongoing about what this means for the history of the parent body. A particularly interesting feature of the angrites is that they are very old, around 4,556 million years, which indicates that asteroidal differentiation and igneous activity was occurring only a few million years after the solar system formed.

**AUBRITES** The aubrites are igneous meteorites that seem to be related to the enstatite chondrites. Unusually for meteorites, the aubrites are a light white–grey colour and have very pale, cream-coloured fusion crusts. This is because the most abundant mineral (75%–90%) is enstatite, a magnesium-rich pyroxene that is also quite abundant in the enstatite chondrites. The most likely formation scenario for aubrites is that they started out like enstatite chondrites, but their parent asteroid was heated to temperatures high enough to cause it to melt and differentiate. Aubrites contain some minerals that are exceedingly rare. An example is found in the aubrite Bustee, which contains centimetre-sized chestnut-brown crystals of a mineral called oldhamite, a calcium sulphide that is not found naturally on Earth.

LEFT The Bustee aubrite, a light-coloured meteorite containing brown oldhamite crystals.

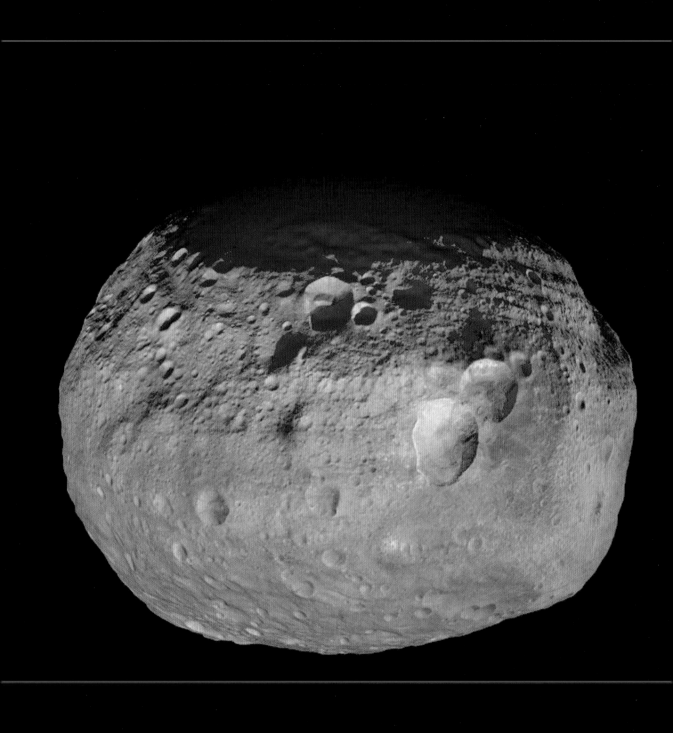

**HOWARDITE, EUCRITE, DIOGENITE (HEDS)** By far the largest group of achondrites, numbering 1971 in January 2018, is the HED group. These meteorites are interesting because they are related to the asteroid 4 Vesta based on spectroscopy (as discussed on p.54).

The eucrite meteorites are basaltic in composition, composed primarily of the silicate minerals pyroxene and plagioclase. The textures of these rocks vary from quite fine grained to relatively coarse grained, indicating that some may have come from the surface of their parent body, while others might have been buried and allowed to cool slowly. These rocks also show evidence of mixing due to impact. Many of the meteorites in this group are breccias, rocks that are broken up and mixed together.

OPPOSITE The asteroid Vesta. This asteroid, or fragments from the same family, are thought to be the parents of the HED meteorites. This image was taken by the Dawn spacecraft.

ABOVE LEFT The Sioux County eucrite, which is thought to have originated on the asteroid 4 Vesta. The sample weighs 153 g (5 ⅓ oz).

LEFT Microscope image of the Pasamonte eucrite showing a basaltic texture. Field of view is 2.5 mm (about ¹⁄₁₀ in) across.

The diogenite meteorites are composed mostly of pyroxene, which when found on its own tells us something about how it formed. Diogenites probably formed in very thick lava flows, where the very slow cooling rate would allow the minerals within the flow to differentiate (similar to how a planet differentiates). When this happens in a lava flow, the result is that layers of minerals are built up. Pyroxene forms first in a thick lava flow and, because it is heavier than the melt, rains out to form a layer on the bottom. Diogenites probably come from this layer of accumulated pyroxene. Finally, the howardite meteorites are breccias and contain bits of both eucrites and diogenites. These probably formed when a large impact excavated and mixed material from the surface and from deep within the parent body.

The HED meteorites, although very different from each other in mineralogy and texture, have chemical similarities that strongly suggest they formed on the same parent body. They also have links to the stony-iron mesosiderites.

BELOW Microscope image of the Johnstown diogenite. Diogenites are coarse grained and composed primarily of one mineral, pyroxene. Field of view is 3.8 mm (about 3/20 in) across.

## NOT CHONDRITE OR ACHONDRITE?

There are several groups of meteorites that fall in between chondrites and achondrites. They are very primitive chemically, but their textures indicate they have experienced heating and sometimes even extensive melting. Four groups of meteorites show these features. These meteorites now form a type called primitive achondrites.

BELOW Microscope image of the Lodran meteorite. This meteorite is the type specimen of the Lodranite meteorites. The lodranites are related to the acaplucoites but lodranites are more coarse-grained. Field of view is 2.5 mm (about $1/10$ in) across.

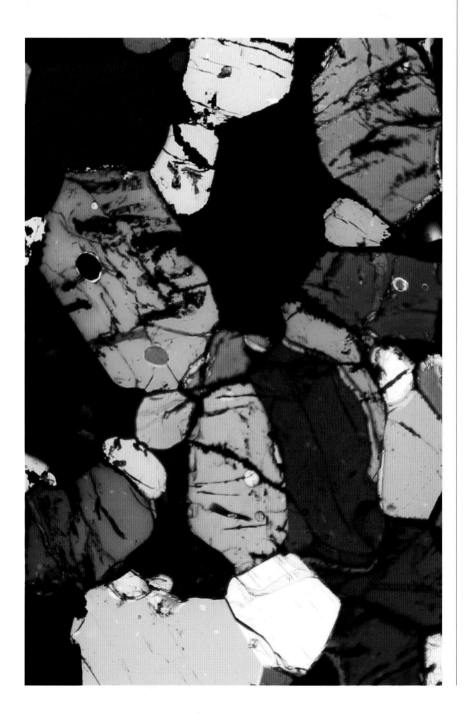

ABOVE Optical microscope image of the Hajmah (a) ureilite. Ureilites contain coarse grained olivine and pyroxene but, unusually, also contain high amounts of carbon in the form of graphite and diamond. Field of view is 5 mm (¹/₅ in).

## BRACHINITES

Brachinites are named after the Brachina meteorite that was found in 1974 in South Australia. This is a small group of meteorites, with only 41 members as of December 2017. They contain mostly olivine, with minor amounts of pyroxene and plagioclase, as well as metal and sulphides. Their textures indicate that they experienced some heating, but it is unclear how much. The question of whether they are primitive or the product of extensive heating remains unanswered.

## UREILITES

Ureilites are one of the oddest and least understood types of meteorites. They are made up of the minerals olivine and pyroxene, but also contain veins of carbon-rich materials in the form of graphite and diamond. Diamonds on Earth mostly form at extreme depths where the pressure is high enough to convert graphitic carbon to diamond. Asteroids are much smaller than the Earth, and thus do not have high enough pressures, even in the centre of the body, to create diamonds.

The pressure comes from impacts in the case of meteorites. Collisions between two asteroids can create very high pressures in a short time. Such events obscure the original nature of the rocks, and unravelling the evolution pathway of these rocks is therefore difficult. How these rocks formed is also a mystery, combining as they do high-temperature olivine and pyroxene with delicate carbonaceous material and volatile gases that must have always remained at a low temperature.

## WINONAITES/IAB IRONS

The winonaites comprise a small group of meteorites that contain mostly low-calcium pyroxene, with smaller amounts of olivine, plagioclase, metal and sulphide. It is thought that these meteorites come from a body where the heating, melting and differentiation began, but was interrupted, most likely by a catastrophic impact. This collision mixed up bits of heated material (the metal) with unheated material to form what we see today.

The IAB iron meteorites form the second largest group of irons, with more than 300 members. Their chemistry suggests that they do not come from a well-formed core of a differentiated asteroid. A very strange thing about these meteorites is that they contain silicate material, but unlike that in pallasites (where it is only olivine) or the mesosiderites. The silicate in these irons is very randomly distributed and is primitive (almost chondritic) in nature. The textures of the silicates show that they have been heated. The silicates found in the IAB irons are related to the stony winonaite meteorite group through oxygen isotopes, mineralogy and mineral chemistry.

BELOW A slab of the IAB iron meteorite Maslyanino showing the size, shapes and distribution of silicate inclusions. The slab is about 10 cm (4 in) wide.

## ACAPULCOITES/LODRANITES

Named after meteorites that fell in Acapulco, Mexico, and in Lodran, Pakistan, this group shows evidence of partial melting of silicate materials. The metallic portion is missing. The acapulcoites have smaller grain sizes than the lodranites. In this grouping, the acapulcoites represent the original composition, but it has been heated to lose its chondritic texture. Very thin veins of sulphide are found along silicate grain boundaries. These form at temperatures just below that which will melt silicate.

The lodranites are very coarse grained, and some of the minerals that are found in the acapulcoites are not found in lodranites. This means these minerals melted and migrated away. The lodranites are called rock residues (the material that is left behind when some of the rock is melted and moves to another part of the asteroid).

These meteorites formed when a chondritic body was heated to a temperature that was able to melt some sulphide (about 1,000°C or 1832°F) and silicate (about 1,100°C or 2,012°F). The acapulcoites did not get as hot as the lodranites, so the temperature in the parent body was not evenly distributed.

# LUNAR METEORITES

The moon is our nearest celestial neighbour, orbiting around the Earth nearly 400,000 km (¼ million miles) away. It is the only astronomical body on which humans have so far set foot and returned samples. For that reason we know more about the moon than any other body apart from the Earth. In addition to rocks returned from space missions, we also have samples of the moon that arrived on our planet by themselves, in the form of meteorites.

BELOW A slice of the lunar meteorite Northwest Africa 482. This meteorite is very rich in the feldspar mineral anorthite, which gives it its pale colour. Meteorites like this are believed to be derived from the Lunar highland regions. The longest edge is about 5 cm (2 in).

The idea that some meteorites may have come from the moon is a very old one – more than three centuries ago some scientists suggested that all meteorites originated from the moon. However, the discovery of asteroids provided a new possible source. Later, results from camera networks showed that the vast majority of meteorites had orbits from asteroids and not from the moon. Then in 1981 an unusual meteorite was found by an American expedition in Antarctica. Allan Hills (ALH) 81005 is dark grey, containing centimetre-sized patches of white material. Once the sample was sent to the USA, it was immediately recognized as looking extremely similar to some of the Apollo samples from the moon. Since ALH 81005 was discovered, more than 300 lunar meteorites have been found. Most of these were also found in Antarctica, but one was found in Australia, and the rest in the Sahara Desert.

The dark and light material present in lunar meteorites corresponds to different minerals. Looking up to the moon on a clear night you can see that there are pale-coloured regions, called highlands (made up mostly of the calcium-rich mineral anorthite), and darker regions called maria (made up of basalt). In lunar meteorites, the dark material consists of the minerals pyroxene, olivine and plagioclase, and the light-coloured material is mostly anorthite. The meteorites were identified as being lunar because the abundances of major and minor elements within them matches both those measured by remote-sensing of the moon and direct analyses of Apollo samples. The oxygen isotopic composition of lunar meteorites is also identical to Apollo samples. The meteorites contain abundant nitrogen and noble gases, which were implanted by the solar wind. This also suggests the meteorites came from a place such as the lunar surface that has no atmosphere to protect it from the wind blown out by the sun.

BELOW A slice of the Lunar meteorite Northwest Africa 773. This meteorite is dark in colour and composed of olivine and pyroxene and some feldspar. Meteorites like this one are believed to have originated in the basaltic, mare regions of the Moon. Length about 4.5 cm (1¾ in).

The lunar meteorites provide much extra information to supplement that from the lunar return samples. The lunar surface area sampled by all the missions to the moon only accounts for around 9% of the lunar surface. The lunar meteorites seem to come from a wider range of different regions, as they have a greater diversity in compositions, and some have originated on the far side of the moon, well away from all the previous sample sites. The areas visited by the Apollo astronauts may not be typical of the moon as a whole, and so samples from different lunar regions provide an extremely welcome addition to our knowledge of lunar geology.

## THE SPACE RACE

ABOVE Early morning view on 9 November 1967 of Pad A, Launch Complex 39, Kennedy Space Centre, showing Apollo 4 Saturn V prior to launch later that day. This was the first launch of the Saturn V.

In the 1950s, the 'cold war' between the USA and USSR had become intense and the two countries were locked in competition with each other. One of the ways they wanted to assert their technical superiority was to become the first country to visit the moon – the Space Race had begun. In the early 1960s, American president J F Kennedy announced his intention to send men to the moon by the end of the decade: 'We go there not because it is easy, but because it is hard,' he said. Both countries invested heavily in their space programmes; for the Americans, this was the Apollo programme, and for the Russians, the Luna programme. In July 1969, the Americans were the first to land on the moon and returned the first samples back to Earth, the first extraterrestrial samples we had apart from meteorites. Ultimately, scientists from all over the world benefited from the political jostling between countries by gaining access to these unique samples.

### THE NEW SPACE RACE
After the political impetus of the Space Race had dissipated there was little to encourage governments to return to the moon. Indeed, in the decades after Apollo, the only major lunar mission was the NASA Clementine mission, which mapped the composition of the surface of the moon. The dawn of the new millennium, however, heralded some new interest in lunar exploration, and this time many more space agencies are involved. Several countries decided to embark on voyages of discovery to our nearest neighbour. These missions include SMART-1 (European Space Agency), Kaguya or Selene (Japan), Chandrayaan-1 (India), Chang'e (China) and Lunar Reconnaissance Orbiter (USA-NASA). The aim of many of these missions is to rediscover space flight, with the objective of later exploring further out in the solar system and in some cases of trying again to take people to other planets. It is to be hoped, in our lifetimes, we will once again see people walk on other worlds.

# THE ORIGIN AND EVOLUTION OF THE MOON

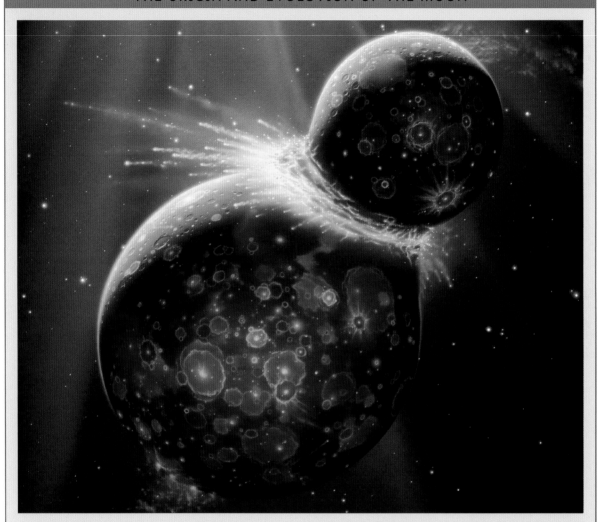

ABOVE The grazing impact of a Mars-size body with the proto-earth more than 4 billion years ago is believed to have led to the formation of our moon.

Samples from the Apollo and Luna missions, and lunar meteorites, have allowed us to understand something of the history of the moon. The moon–Earth system is unique in the solar system because the size of the moon is so large compared to the size of the Earth. Rocks from the moon and Earth show some clear similarities but also important differences. The oxygen isotope compositions of the Earth and moon are indistinguishable, pointing to a common origin for both. However, the moon has a lot less iron than the Earth, it has no large iron-rich core, and it has a much lower abundance of volatile elements. In particular, it has hardly any water, in great contrast, to the watery Earth. In the 1980s, scientists brought all this together to produce a theory, called the Giant Impact Theory and now widely accepted, of how the moon formed. A planet the size of

Mars, called Theia, hit the young Earth at an oblique angle. This completely melted most of both planets, and forced a ring of material to spin out of the Earth, which accreted quickly into the moon. The moon was initially extremely hot, and the outermost parts were molten, forming a 'global magma ocean', a sea of liquid rock over the whole surface. As this ocean solidified, it formed a layered structure, with the lighter anorthosite rock floating to the top. The moon has been bombarded by asteroids, leaving many craters over its surface. Some of the giant impact basins were then filled with lava, during episodes of basaltic volcanism; these are the dark circles you can see on the surface of the moon.

# MARTIAN METEORITES

There is a group of meteorites with features that indicate they did not form on an asteroid. All the members of this group have young ages, and all but two are between about 1,300 and 150 million years old. They also have mostly igneous textures that could only crystallize from a melt (or in the case of one unusual sample, contain fragments of igneous rock). The young ages imply that these rocks come from a body that has been geologically active – one that has been hot, or even molten, for a long time. Even melted asteroids cooled off by about 4,000 million years ago because they were too small to stay hot. For rocks to be as young as about 150 million years old means that these came from a planet-sized body. But which planet? There are several lines of evidence that show these rocks came from Mars.

## HOW DO WE KNOW THESE METEORITES ARE FROM MARS?

The Martian meteorites were known to be different from asteroidal meteorites long before they were determined to be from Mars. The lines of evidence that allowed scientists to deduce their origin are as follows. First, the textures of the vast majority of the rocks are igneous. This means they formed from melted rock. There are asteroidal meteorites that formed from melted rock, so this does not

LEFT The planet Mars with the Valles Marineris trough system visible in the foreground.

definitively prove they are from Mars. Second, although these rocks contain the same minerals as found in asteroidal meteorites, there are differences in the ratios of some elements. The iron to manganese ratios in the olivine and pyroxene minerals, together with the oxygen isotopic composition, showed that the rocks are related to each other and came from the same parent body. Third, their ages indicate they come from a planet-sized body. The ages of all but two of these meteorites are less than 1,300 million years. The 'clocks' that record ages in meteorites start when the rock cools down from the molten state.

All asteroids are too small to be hot for very long, and that is why even the igneous-textured asteroidal meteorites are all 4,000 million years old or older. Young ages indicate a large body that can keep melting rocks for longer, so that means a planet. The final piece of evidence that proves these rocks are from Mars comes from the composition of gases that were trapped in the Elephant Moraine (EET) A79001 meteorite (and have since been found in other meteorites of this type). This meteorite contains black glass in addition to the other igneous minerals. The glass formed by a shock melting, possibly during the impact that launched it from the surface of Mars. Analysis of the gas trapped in the glass shows that it is identical to the composition of the 'air' on the surface of Mars, as measured by the Viking landers in 1976. Because all the rocks are related through their mineral chemistry, ages and textures, they all must come from Mars.

BELOW Iron to Manganese ratios for meteorites from different bodies.

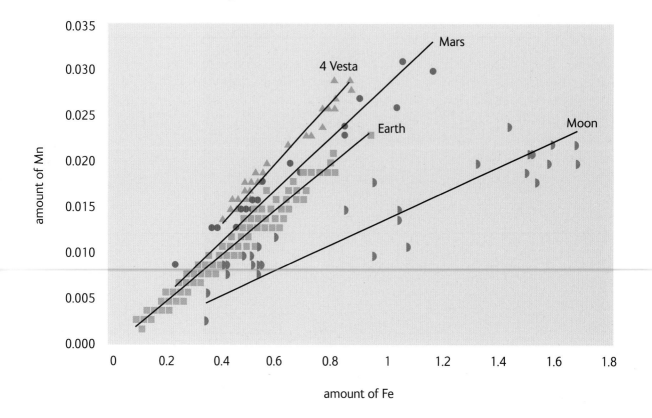

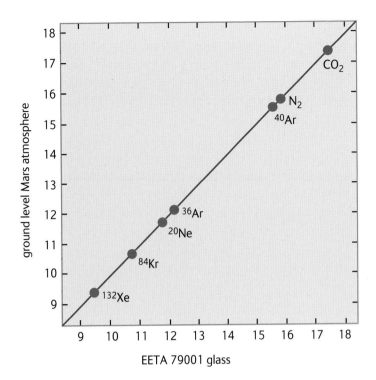

Rocks are launched from Mars by the process of impact ejection. As we have discussed (Chapter 3), impacts and collisions between planetary bodies are important processes. If an asteroid or comet were to hit the surface of Mars at a small enough angle and with enough speed, then ejecta from the impact could escape the gravity of Mars to orbit the sun, and some of this material can land on the Earth as meteorites. Meteorites from Mars are very rare. As of January 2018 there are 200 specimens from more than 100 individual meteorites. Of the Martian meteorites in the world's collections, five are falls (see Chapter 1). These are Chassigny, France (1815), Shergotty, India (1865), Nakhla, Egypt (1911), Zagami, Nigeria (1962) and Tissint, Morocco (2011). Interestingly, Chassigny and Zagami fell on the same day – 3 October – but 147 years apart. Most of the rest have been found in the hot deserts of Africa or Oman (164) and the cold desert of Antarctica (29). The remaining meteorites were found in Brazil and the USA.

Martian meteorites were originally classified into one of three groups, named after three of the best-known meteorites of this type – Shergotty, Nakhla and Chassigny (SNCs). Shergotty is a broadly basaltic rock composed mostly of the minerals pyroxene and plagioclase. The rocks that resemble Shergotty are called shergottites. Nakhla is classified as a clinopyroxenite, which means that these types of rocks formed in the middle of a thick lava flow or a sill and represent something from a shallow depth. The rocks that resemble Nakhla are called nakhlites. Chassigny is of a very rare rock type, called a dunite, meaning it is composed almost entirely of olivine. Until very recently, no other Martian meteorites had dunitic features. In 1994 the meteorite ALH 84001 was shown to have a Martian origin. This meteorite is very different from the other meteorites, being extremely old at 4,100 million years, and composed of orthopyroxene. In 2011 a very unusual sample was described. It was very dark in appearance and showed a distinctive brecciated texture, with different fragments of igneous and impact rocks and minerals. This unusual rock was nicknamed 'Black Beauty' by the scientists who classified it, but its official name is NWA 7034. Like ALH 84001, this meteorite did not fit into the classic SNC scheme and so meteorites from Mars are now called Martian meteorites.

## SHERGOTTITES

Shergottites are silicate rocks that are now divided into four subgroups. One group, the basaltic shergottites, consists of fine-grained rocks composed of equal amounts of pyroxene and plagioclase. These rocks formed in a lava flow on the surface of Mars. The other groups are more coarse grained, which means they cooled more slowly. They probably formed deeper in the Martian crust than the basaltic rocks. All of these meteorites are heavily shocked but not brecciated. The shergottites erupted onto the Martian surface between about 150 and 575 million years ago. Most of the shergottite rocks we have on Earth were launched from Mars in two different impact events between about 0.5 and 3 million years ago, although one sample (Dhofar 019) has an 'ejection' age of 20 million years.

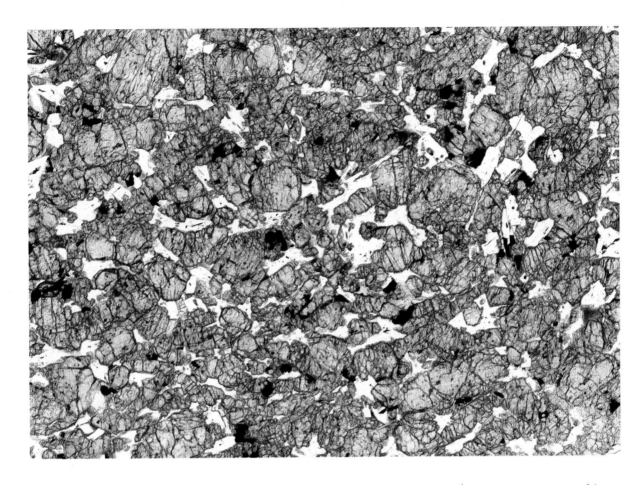

ABOVE Microscope image of the Zagami shergottite. The fractures in the pyroxene mineral grains and the paler patches of glass show that the rock has been shocked. Field of view is 5 mm (¹⁄₅ in).

LEFT Shergottite Sayh al Uhaymir 008, found in Oman in 1999. The specimen is about 10 cm (4 in) long. A piece of this meteorite is actually returning to Mars, as a calibration sample for the SHERLOC instrument on board the Mars 2020 mission.

## ASTEROIDAL METEORITES ON MARS

Given the cratering record on Mars, it is not unexpected that asteroidal meteorites should be on the surface of Mars too. The Mars Exploration Rovers (Spirit, Opportunity and Curiosity) that have been roaming around Mars since January 2004 have found such meteorites. The most recent Mars Rover, Mars Science Laboratory (Curiosity), has also found a number of iron meteorites during its travels in the Gale crater.

ABOVE A composite of high-resolution images from the ChemCam (white outlines) and MastCam instruments on board the Mars Science Laboratory Rover. More commonly known as Curiosity, the rover discovered these two iron meteorites in 2014. The larger meteorite (rear) is called Lebanon and is approximately 2 m (7 ft) wide.

### NAKHLITES

Nakhlites are named after Nakhla, a meteorite that fell in Egypt in 1911. These meteorites are clinopyroxenites – they are composed almost entirely of the green mineral augite with some minor amounts of the mineral olivine. They also contain small amounts of minerals formed in the presence of water, such as clays, carbonates and sulphates. These formed in lava flows at or very near the Martian surface and were erupted from a volcano over a period of about 100 million years, much longer than for any known volcanoes on Earth. These meteorites are older than the shergottites, with ages of around 1,300 million years. They were also launched from the surface of Mars before the shergottites, roughly 11 million years ago. There is a famous story surrounding the Nakhla meteorite fall. There was a witness statement that a dog had been hit by one stone '.....and left it like ashes in the moment'. It is now believed that this is a misinterpretation of the Arabic and describes the fact that the dog was not bothered by the meteorite! The few reports of meteorites that have been recovered almost immediately after landing are described as being cold or slightly warm, so it is impossible that the dog was incinerated by a piece of the Nakhla meteorite.

The Nakhla meteorite fell as a shower of stones in Egypt in 1911. Although one of the stones is rumoured to have killed a dog, this story is almost certainly not true.

Microscope image of the Nakhla meteorite showing large pyroxene crystals. Field of view is 5 mm (¹⁄₅ in).

ABOVE Microscope image of the Chassigny meteorite showing shocked and deformed olivine grains. Field of view is 5 mm (1/5 in).

## CHASSIGNITES

Until very recently there was only one member of the chassignite group, the meteorite Chassigny. A further two meteorites have now been added to this group. These meteorites are dominated by the mineral olivine and are called dunites. This mineral is relatively dense (because it contains about 28% iron and magnesium) and when it forms will sink within the melted rock as it crystallizes. These rocks most likely formed at depth within the Martian surface. They have similar formation ages to the nakhlites, and other chemical and mineralogical evidence suggests that they may be related.

## ORTHOPYROXENITE

ALH 84001 is the only member of this subgroup. It was discovered in Antarctica in 1984 and was originally classified as a diogenite (see p.81), based on its mineralogy. It was not shown to be Martian in origin until 1994. ALH 84001 is composed almost entirely of orthopyroxene, making it very different from other meteorites from Mars. It is the oldest of the Martian meteorites, at 4,100 million years, and its history is complicated by shock and thermal metamorphism. Interestingly, it contains carbonate minerals. On Earth the presence of carbonate minerals generally means a rock has come into contact with liquid water at some point in its history. The carbonates in ALH 84001, however, are complex in both their chemistry and shape, which has led to several theories about how they formed. Some think they formed at low temperature, and others that they formed at high temperature. Still others say they formed during the impact event that shocked the silicate minerals. The consensus seems to be that the carbonates record a varied history of impact-driven interaction between fluid and rocks.

LEFT A piece of the ALH 84001 orthopyroxenite. It is about 5 cm (2 in) across.

BELOW Patches of orange carbonate in ALH 84001. The field of view is about 3 mm ($^1/_{10}$ in).

## ARE THERE FOSSILS IN ALH 84001?

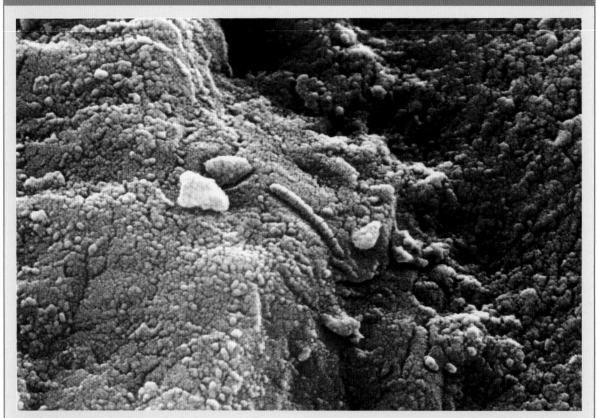

The ALH 84001 meteorite was thrown into the limelight in 1996 when scientists from NASA's Johnson Space Center, USA reported finding evidence in the rock consistent with it containing fossils. Throughout ALH 84001 there are patches of bright orange carbonates up to a few millimetres across. A team of scientists studied the meteorite using a variety of techniques, and found tiny structures in the carbonates that looked like fossilized bacteria. Associated with the carbonates were organic compounds, which could have confirmed that the tiny structures were biological in origin. However, these have been proved to be contamination from the Earth. In spite of that, the report of possible fossils in a meteorite has triggered a whole new area of research called astrobiology, dedicated to understanding the interactions between biology and planetary geology. Astrobiology brings together biologists, chemists and planetary scientists to learn about whether life can exist outside Earth. The investigations of ALH 84001 and the development of astrobiology have revitalized the Mars Exploration Program which, among other things, is attempting to answer the question of whether there are

ABOVE Scanning electron microscope image (size about 1 μm) of the surface of the orange carbonates in ALH 84001, showing what was termed a 'microfossil'.

habitable environments on Mars. This has led to the launch of the Mars Exploration Rovers (MERs) and other missions (there are currently four missions in orbit or on the ground at Mars, with several more in the planning stages). While out traversing the surface of Mars, one of the rovers found a rock that was named Bounce Rock. This rock is chemically similar to EET A79001, the rock from which the gas that matched the atmosphere of Mars was extracted and thus indirectly ALH 84001, and helped prove that this group of meteorites on Earth was from Mars. Future exploration plans for Mars include ESA's ExoMars 2020 which will look for life. It will target the oldest Martian surfaces, as it is in the earliest history of Mars that the planet would have been most likely to be habitable, more wet and warm than it is today. A rover will traverse the surface and search for organic molecules, and it will drill down up to 2 m (6½ ft) to find material protected from irradiation on the surface.

## BASALTIC BRECCIA

NWA 7034 was the first recognized Martian meteorite with a complex brecciated texture. Found in Morocco in 2011, it is composed of fragments or 'clasts' of rocks and minerals that have formed in multiple, different igneous and impact processes. Interestingly, it also contains the most water of any of the Martian meteorites, about 0.6%. The chemical composition of NWA 7034 is similar to the rocks and soils measured by NASA's Mars Exploration Rover 'Spirit' at Gusev crater, and also to those measured in the Martian crust, as measured remotely by the Mars Odyssey mission. NWA 7034 is the second-oldest Martian meteorite, with an age of about 2,100 million years; however, this age is debated as it is a 'whole-rock' age, meaning that the age is an 'average' of all the different fragments, each of which may well have a different age depending on when they formed. Some of the igneous clasts in this type of meteorite contain a mineral called zircon and these zircons are very old, dating back to about 4,430 million years. This implies that the planet Mars formed a crust about the same time as the Earth and the moon - these zircons are the tiny remnants of that original Martian surface. The complex brecciated texture indicates that this rock type has had a complex history since its initial formation, the different clasts recording a wide array of igneous and impact events. By January 2018 a further nine Martian basaltic breccias had been found in northwestern Africa.

LEFT A virtual slice through NWA 11220, a Martian basaltic breccia paired to NWA 7034. This image is made from X-ray computed tomography data and shows the complex brecciated texture of the rock. The specimen is about 2.5 cm (1 in) across.

# CHAPTER 5

# Comets

T HE VAST MAJORITY OF METEORITES come from asteroids (99.4%), a few come from the moon (0.4%) and a few from the planet Mars (0.2%). A very rare type of unmelted meteorite, the carbonaceous Ivuna type (CI) chondrites – of which only nine are known (< 0.03% of known meteorites) – have characteristics that suggest they may have originated from comets.

Comets are very important constituents of the solar system and provide scientists with evidence of the chemical and environmental conditions that existed during the very earliest stages of solar system formation. They must have formed beyond the nebula 'frost' or 'snow' line, where the temperature was low enough for water ice to form. Comets are the preserved remnants from these very cold, ice-bearing regions that composed 99% of the solar nebula disc from which our solar system formed. They are mixtures of both nebular and pre-solar materials that accreted over a vast area of the very young solar system, and thus they record the different chemical, temperature and physical environments over a vast area of space.

OPPOSITE Comet Hale-Bopp which appeared in the night skies of the northern hemisphere in 1997.

## METEOR SHOWERS

As we learned in Chapter 1, approximately 40,000–60,000 t of extraterrestrial material lands on Earth every year, the majority of which is in the form of tiny dust grains usually less than 1 mm (1/25 in) in size; importantly, most of this dust is believed to originate from comets.

Every time a comet approaches the sun it loses material; the ices melt and vaporize, releasing trapped dust and larger particles. Typically a comet loses up to 0.1% of its mass each time it passes close to the sun. Calculations showed that Comet Halley lost about 500 million t during its last orbit around the sun. Based on this amount of mass loss, Comet Halley will only survive for another 2,200 orbits, or another 167,000 years or so. In fact, it is unlikely that Comet Halley will last for even this relatively short amount of time;

ABOVE An impression of the Leonid meteor shower of November 1799, so-called because the meteors appear to radiate from the constellation Leo.

RIGHT A transmission electron microscope image (TEM) of a cometary particle from the Grigg-Skjellerup comet. This particle was collected by a high-altitude NASA aircraft.

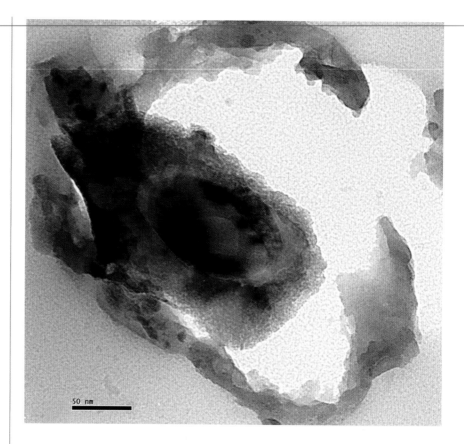

50 nm

most comets break up rather than shed mass linearly, so it is likely to fragment long before it 'runs out of steam'.

Direct evidence from spacecraft such as Giotto and Stardust, as well as observations of meteors, indicate that particles of 1 mm (¹⁄₂₅ in) and larger can make up a large proportion of the material ejected from the comet. Sometimes very large particles and fragments of 1 m (3¼ ft), or larger, may also be ejected. Over time, gravity and solar radiation smear out this material so that eventually it occupies the whole of the comet's orbit around the sun. In some cases, the Earth's orbit will coincide with a cometary orbit. When the Earth encounters this cometary debris it produces a meteor shower as the larger particles burn up in the atmosphere; no material hits the ground as a meteorite. However, very tiny dust grains, usually less than 1 mm (¹⁄₂₅ in) in size, can reach Earth's surface; these do not produce meteors, as they are too small, and they simply 'float' down through the atmosphere. Interestingly, some particles from the Comet Grigg–Skjellerup have been collected by NASA's scientific aircraft flying at very high altitudes.

Every year there are several meteor showers, some of which can produce impressive meteors easily visible to the naked eye. Meteor showers are named after the stellar constellation from which, or near to which, they appear to originate. For example, the Orionids appear to radiate from the constellation of Orion and the Geminids from the constellation of Gemini.

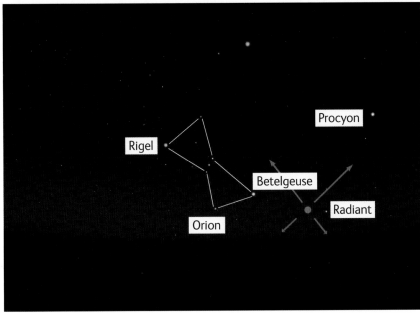

ABOVE AND LEFT Illustrations showing the radiants of the Geminid and Orionid meteors, so named as they appear to originate near to these constellations.

As mentioned above, although the dates of the showers are predictable, the intensity (number of meteors seen, brightness, etc.) can be highly variable. This is because the intensity depends on where the comet is in its journey around the sun and whether the stream has just been 'recharged' with debris from a recent passage. For example, the Leonid shower, which occurs in mid-November, is related to the Comet 55P/Tempel–Tuttle, which has a 33-year orbital period.

Historical records show that there were spectacular storms in 1833 (when hundreds of thousands of meteors per hour were reported) and in 1866 and 1867. However, the shower of 1899 was a disappointment and another spectacular storm was not observed until 1966. Astronomers predicted that the shower of 1999 would be similarly stunning, and over 3,000 meteors per hour were recorded in some locations.

## COMET SHOWERS

| Shower name | Date | Approx. no. meteors per hour at maximum | Parent comet |
| --- | --- | --- | --- |
| Quadrantids | early January | 100 | 2003 EH1 |
| Lyrids | second half of April | 10 | C/Thatcher |
| Eta Aquarids | late April to mid- May | 40–85 | 1P/Halley |
| Southern Delta Aquarids | late July | 20 | 96P/Machholz |
| Perseids | mid-August | 60 | 109P/Swift–Tuttle |
| Draconids | early October | 10 | 21P/Giocobini–Zinner |
| Orionids | late October | 25 | 1P/Halley |
| Leonids | mid-November | 20–40 | 55P/Tempel–Tuttle |
| Geminids | mid-December | 120 | 3200 Phaethon |

Some of the major meteor showers visible to the naked eye are shown in the table above. While the date range for the meteor showers is predictable, the date of maximum intensity and the number of meteors seen is variable. Similarly, some showers are visible from only some parts of the world. If you are interested in finding out the dates for any showers visible in your location, local astronomical societies will be able to provide more detailed information.

Predictions of meteor showers can be complicated because they evolve over time due to interactions with solar rays and gravitational effects of planets and moons.

Other planets and also moons are likely to have meteor showers, as long as their atmospheres are not too dense. Meteors impacting onto the surface of the Moon can sometimes be observed from Earth. A meteor on Mars was also observed by the NASA rover Spirit.

OPPOSITE An image of the long period comet C/2011 W3 (Lovejoy) taken from Scotland by Colin Legg. It is called a sun grazer comet as its orbit took it so close to the sun in December 2011.

# COMETS THROUGHOUT HISTORY

Throughout history comets have fascinated and often frightened peoples around the world. The appearance of a comet was frequently seen as a 'message from the gods' or a portent, predicting (usually negatively) major events such as wars, disease or the death of a leader. After the appearance of a comet in about the year 60, the Roman populace assumed that this meant bad news for eminent people, especially the most eminent, the Emperor Nero. On the advice of an astrologer,

BELOW Chinese astronomers recorded detailed observations of comets and other celestial phenomena.

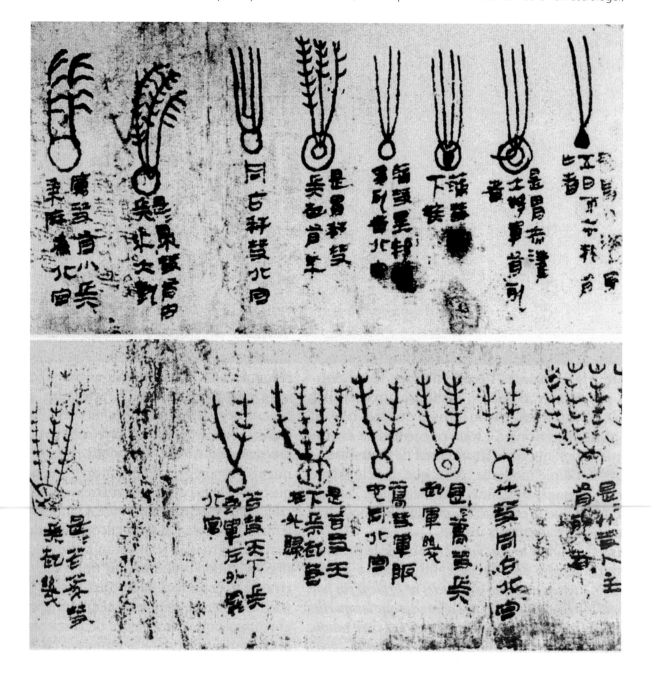

Nero decided to massacre a large number of the Roman nobility as this was a good way to 'deflect' the wrath of the gods. Of course, one may presume that it could also be a 'good way' to get rid of any potential rivals to the throne! Nero survived and the Roman citizens' cometary-induced superstitions were confirmed by the deaths of many important people.

LEFT The painting 'Adoration of the Magi' depicts Comet Halley as the 'Star of Bethlehem'. In 1985 the European Space Agency sent a mission to study Comet Halley, which was named 'Giotto' after the painter of this picture, Giotto di Bondone (1267–1337).

BELOW The appearance of Comet Halley in 1066 was seen as an omen and recorded in the Bayeux Tapestry (see top right-hand corner).

ABOVE Comet 1P/Halley taken 8 March 1986 Easter Island. Comet 1P/Halley is a short-period comet, whose orbit around the sun makes it visible to Earth every 75 years or so. Its last 'visit' was in 1986 and its next projected return is in 2061 and so it is likely that readers of this book will see it at least once in their lifetimes.

Chinese astronomers also took the observation and recording of comets very seriously. Observations of astronomical phenomena were of great interest to the emperors of China, who viewed the appearance of comets, eclipses and other phenomena with great superstition. The detailed records of Chinese astronomers include the observations in 1054 of a 'new star', which was actually a stellar explosion or supernova (the remnants of which can still be seen today as the Crab Nebula), and also the appearance of Comet Halley in the year 837. Comet Halley also appears in two famous European artistic works, the Bayeux Tapestry and Giotto's painting 'Adoration of the Magi', both of which depict important historical events – the Norman conquest of England in 1066 and the birth of Jesus Christ.

The word comet is derived from the Greek word 'kome', which means 'hair of the head'. Aristotle, the famous Greek natural philosopher, first described comets as 'aster kometes', meaning 'stars with (long) hair', which was later shortened to 'kometes'. It is not surprising that Aristotle chose this word to describe his observations, as the most notable feature of a comet is the long tail that streams away from the bright head or nucleus.

# WHERE DO COMETS COME FROM?

Comets are classified into two types on the basis of the length of time it takes to complete an orbit of the sun, known as the comet's period. Long-period (LP) comets take more than 200 years to orbit the sun and short-period (SP) less than 200 years. SP comets can be further subdivided into Halley-type comets (periods of 30–200 years) and Jupiter-family comets (periods of less than 30 years). Jupiter-family comets are so named because the furthest part of their orbit from the sun (the aphelion) reaches to the orbit of Jupiter.

Comets are unstable near the sun, and where their orbits can be disrupted by the gravitational action of planets. There are two major reservoirs where comets can survive and are 'stored' without the detrimental effects of solar heating or gravitational perturbations. LP comets have elliptical orbits and approach the sun from random directions. The main reservoir for LP comets is the Oort Cloud, which is believed to exist at the outermost edge of the solar system, well beyond the orbit of Pluto. As it is so very far away, comets in the Oort Cloud cannot be observed by telescopes. However, theoretical calculations by the Dutch astronomer Jan Hendrik Oort suggested that it may be the source region for LP comets.

SP comets approach the sun at fairly shallow angles and were once thought to be LP comets that had been perturbed by the gravitational fields of the giant planets into SP orbits. It is now known that SP comets actually originate in their own reservoir, the Kuiper Belt (or Edgeworth–Kuiper Belt). As with the Oort Cloud, the existence of the Kuiper Belt was first proposed on the basis of theoretical

BELOW Image of a binary Kuiper Belt object 1998 WW31 taken by the Hubble Space Telescope. The image is made up of six different exposures which show the orbit of the smaller object relative to the larger. The smaller body is estimated to be about 130 km (90 miles) in diameter and the larger body about 150 km (93 miles).

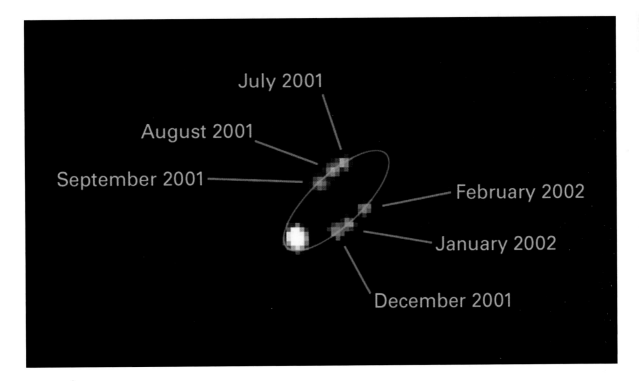

July 2001

August 2001

September 2001

February 2002

January 2002

December 2001

# IS PLUTO A PLANET?

Secha

Quaoar

Eris

Pluto

Moon

Earth

In 2006, the 2,500 astronomers at the annual meeting of the International Astronomical Union (IAU) voted to reclassify Pluto as a 'dwarf planet'. Pluto was discovered in 1930 by the American astronomer Clyde Tombaugh and, prior to 2006, was considered to be the ninth and outermost planet of the solar system. However, its classification as a planet has been a matter of debate since Tombaugh's discovery. Pluto has some rather unusual characteristics for a 'traditional' planet. Its orbit is highly tilted compared with the other planets. Also, its orbit is highly eccentric: at times it is closer to the sun than its neighbour Neptune. Finally, it is considerably smaller than the other planets, especially in comparison to the outer planets to which it is closest. Indeed, our own moon and

ABOVE Illustration showing the relative sizes of the Earth, moon and TNOs and KBOs, including Pluto, which is no longer designated as a planet.

some of the moons of Jupiter and Saturn are larger than Pluto. Pluto's relegation to dwarf planet status was on the cards since the early 1990s, with the first observations of large Kuiper Belt or Trans-Neptunian Objects. Over the last decades it has become clear that Pluto is one of a large family of bodies in orbit around Neptune and beyond. New rules for planets were devised by the IAU: they have to be round in shape, to be orbiting a star and to have swept away all the material within their orbit. Pluto fails on this last criterion and was demoted to be a dwarf planet.

calculations of SP comet orbits. The Kuiper Belt exists in a region beyond the orbit of Neptune, at a distance of between 30 and 50 AU (1 AU, or astronomical unit, is the average distance from the Earth to the sun, or about 150 million km (93 million miles). With technical advances in telescope design and capabilities the first Kuiper Belt object was discovered by astronomers in 1992; it was an object with a diameter of about 320 km (199 miles) at a distance of 44 AU. We now estimate that there are trillions of objects in the Kuiper Belt. Some of these are being explored by the NASA New Horizons mission.

It should be noted that there is some controversy over the naming of the Kuiper Belt. Its existence was first suggested in 1930 by Frederick C. Leonard, an American astronomer who, coincidentally, also had a great interest in meteorites, and was a founding member of the Meteoritical Society. In 1949 the British astronomer Kenneth Edgeworth discussed the possibility that, in the region beyond Neptune, very early solar system materials would condense into many small bodies rather than planet-sized bodies. A couple of years later, in 1951, the Dutch astronomer Gerard Kuiper proposed that bodies no longer existed in this region as they would have been thrown out of the solar system by the influence of Pluto's gravity (at the time Pluto was thought to be the same size as the Earth). Over the following years many other astronomers joined the debate with variations of the hypothesis describing a belt of material existing beyond Neptune's orbit. Today, many astronomers use the term Trans-Neptunian Object to describe the objects existing in the very outer regions of the solar system.

## COMETARY NUCLEI

In the 1950s, the British astronomer Fred Whipple coined the term 'dirty snowball' to describe a comet's nucleus, and as an approximation this is still a good description. From observations by spacecraft and powerful telescopes we know that the nucleus of a comet is composed primarily of ices with rock, dust and also organic materials. Water ice is the major component of a comet's nucleus, but other frozen compounds are also present including carbon dioxide, methane, carbon monoxide and ammonia. The rock and dust portion is composed of silicates such as olivine and pyroxene, many of which lack a crystalline structure, so are termed amorphous. Up to around 20% of the nucleus may be made up of organic compounds like ethanol, formaldehyde and methanol. The dust grains are coated with the organic compounds and mixed with the water and other volatile compound ices to form a highly porous body with voids and cracks from which jets of gas can escape. In 1985 the European Space Agency's Giotto mission visited Comet Halley and captured, for the first time, high-resolution images of a comet's nucleus. These images showed the nucleus had a very irregular shape. Irregular shapes are extremely common and probably result from a combination of factors including the low mass (and therefore gravity) of the nucleus, the loss

RIGHT An image of Comet Halley's nucleus taken by the Giotto spacecraft.

BELOW A false-colour composite image of Comet Borrelly's nucleus taken by the NASA Deep Space 1 spacecraft. The image shows jets escaping from the nucleus and the coma surrounding it.

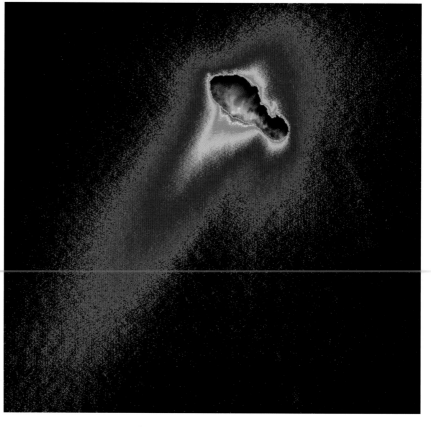

of volatile gases and fragmentation of the nucleus during previous orbits. Further observations have shown that comet nuclei can range in size from a few hundred metres to over 40 km (25 miles) in diameter.

With an object that is composed primarily of ice you may expect the surface of the nucleus to be very bright and shiny. However, the surface is in fact very dark, and comet nuclei are some of the darkest objects that are observed in our solar system. Cometary nuclei reflect less than 4% of the sunlight that falls on them; in comparison a full moon reflects about 15%. The nuclei may be coated with a dark, tar-like substance, formed through the cosmic irradiation of the organic compounds.

## COMETARY TAILS

Perhaps the most characteristic and spectacular feature of a comet is its tail. In fact, a comet actually has two tails, one made from dust and the other – the ion or plasma tail – from charged particles. Both tails point away from the sun, as the dust and charged particles are pushed away by solar radiation pressure.

As a comet approaches the sun the dark-crusted nucleus is heated, and the ices start to vaporize and release gas, dust and rock fragments into space. This escaping material forms a weak atmosphere around the nucleus termed the coma, which can be up to 10 million km (6,250 million miles) in size. A coma will appear when the comet is between 3 and 5 AU from the sun. Fragments from the coma of Comet Wild 2 were captured by the NASA Stardust mission in 2004 and brought back to Earth for study. They showed that the dust grains are composed of a mixture of low- and high-temperature minerals.

The gases released from the comet become ionized by the effect of solar ultraviolet radiation and are swept away by the solar wind at speeds of between 300 and 500 km/s (around 670,000 and 1,200,000 mph), forming the ion tail. The ion tail is usually bluish in colour and is long and straight, stretching over 100 million km (62,100 million miles) in space. Magnetic fields within the ion tail, and their interaction with the solar wind, result in complex structures being formed, which are often observed as lines or ridges in the tail. In some cases the interactions of the magnetic fields can become so intense that the ion tail is 'ripped off' the comet. This happened with Comet Encke in April 2007.

In contrast to the ion tail, the dust tail looks off-white, pinkish or yellowish in colour, caused by sunlight reflecting off the dust grains. The dust tail appears curved, as a result of the individual dust grains being in the same orbit around the sun as the comet. Like the ion tail, the dust tail can reach over 100 million km (about 62 million miles) in length. The grains that make up the dust tail are usually very small, less than 1 μm in size, and are coated with organic compounds. These dust particles are pushed into space, at speeds of several kilometres per second, by the expanding gases that are released from the comet as the nucleus heats up.

# STUDYING COMETS – SPACE MISSIONS AND TELESCOPES

The properties of comets have been measured using ground-based telescopes and dedicated spacecraft, and by analyzing cometary dust particles collected by high-altitude aircraft. Beginning in the early 1970s, telescopes analyzing the infrared spectra of comets have shown the presence of tiny grains of silicate minerals. Mass spectrometers on the Giotto mission to Comet Halley showed that, on the basis of chemical composition, there were three main types of dust present, dominated by: (1) carbon, hydrogen, oxygen and nitrogen (CHON); (2) silicate (silicon, magnesium, iron and oxygen); and (3) mixed grains with CHON and silicate compositions. In 1999, NASA launched the Stardust mission to intercept Comet Wild 2 in January 2004. The spacecraft travelled over 4 billion km (2,500 billion miles) to collect cometary dust particles in its aerogel collector and return them safely to Earth in January 2006. Analyses of the collected particles by hundreds of scientists from around the world have revealed some exciting and unexpected results. Much of the silicate dust that was collected was crystalline; this was surprising, as astronomical observations suggested that the silicates should be mainly glassy or amorphous. Perhaps one of the most remarkable results was the presence of grains that were very similar in composition to calcium and aluminium-rich inclusions or CAIs (see Chapter 4). CAIs were formed in very high temperatures, in the inner parts of the solar system, so it was very surprising to find one in a comet, which formed

LEFT Engineers investigating the Stardust spacecraft prior to its launch in 1999. The white conical-shaped object at the top is the sample return capsule that returned the cometary particles to Earth in January 2006.

ABOVE An image of the nucleus of the Wild 2 comet, taken during a close approach by the Stardust spacecraft. The nucleus is about 5 km (3 miles) in diameter.

BELOW A cometary particle track collected in aerogel. The particle collected from this track has been named 'Inti' and has a similar composition to CAIs.

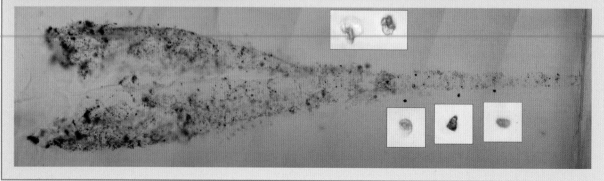

LEFT Comet 67P/Churyumov-Gerasimenko, the target of the Rosetta space mission. The complex shape of this object suggests it is actually formed from two bodies that collided with each other and stuck together.

BELOW AND BOTTOM Images taken by the Deep Impact spacecraft at the time of projectile impact and 16 seconds afterwards – you can see a plume of ejected material in the latter.

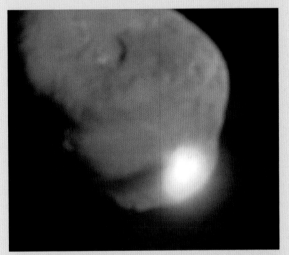

in the cold, outer regions of the solar system. This one result has shown that the young solar system was a very dynamic place, with large-scale mixing of material from the inner to the outer solar system.

In 2005, NASA launched a further cometary mission, Deep Impact, to Comet Tempel 1. This mission was designed both to take images of the comet's nucleus and also to launch a projectile into the surface. The aim was to make a crater that could be mapped and imaged by the spacecraft to determine the interior composition of the nucleus. The projectile successfully impacted the nucleus, but the large amount of dust that was produced during the impact obscured the view of the impact crater (though the Stardust spacecraft is now being used to revisit Tempel 1 to view the crater). Results showed that the nucleus was dustier than expected and also that it was extremely porous, with about 75% empty space.

In 2004, the European Space Agency (ESA) launched the Rosetta mission, which took 10 years to travel 790 million km (491 million miles) to intercept Comet 67P/Churyumov–Gerasimenko. The mission is, to date, the most complex ever designed to study the make-up of a comet. The orbiter took high-resolution images, and instruments will measure the composition of the coma, tail and nucleus. The images beamed back showed that the comet has a 'rubber duck'-like shape, formed as two rounded bodies gently collided. In 2014 the ESA Rosetta space mission became the first to touch down on the nucleus of a comet when its lander, Philae, made

a bumpy landing onto the surface of the irregularly shaped nucleus. The lander survived only long enough to beam back a few measurements, but in 2015 briefly woke up to work some more before stopping forever. The mission showed that comets contain many complex organic molecules such as the precursors of sugar.

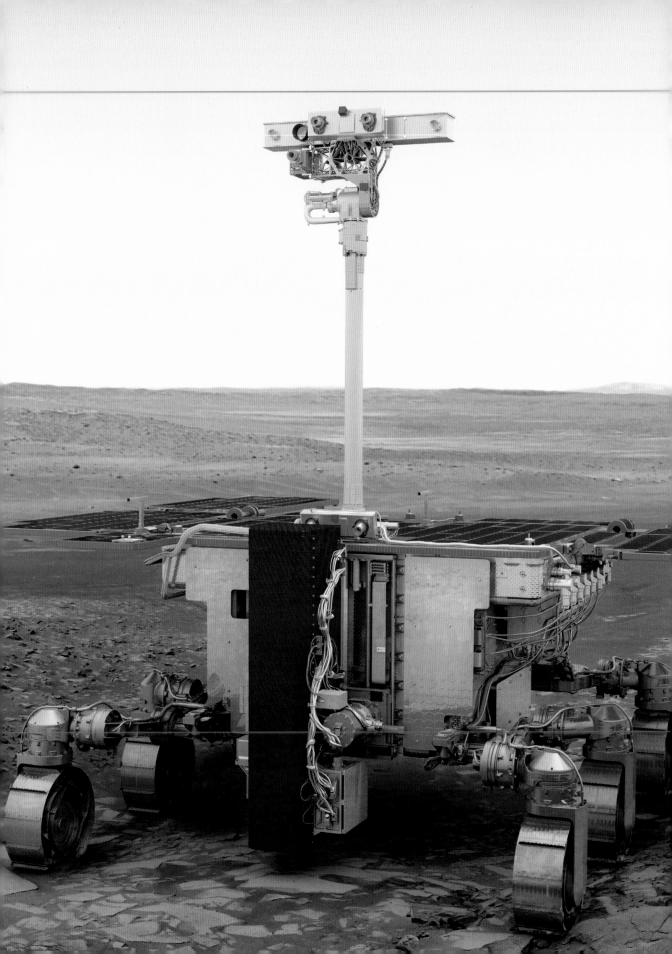

# CHAPTER 6

# The future

I N THE 200 YEARS OR SO SINCE METEORITES became the objects of serious scientific investigation, scientists have gathered a wealth of information to help us better understand how our solar system, and the bodies within it, formed and evolved to their present state. We recognize that meteorites that fall to Earth come from the asteroid belt, the moon, Mars and perhaps even comets. We've learned that impacts and collisions between comets, asteroids and planets within the solar system have played a crucial part in the geological and, in the case of Earth, the biological evolution of those bodies.

Through the study of the unmelted, chondritic meteorites we can date the 'birth' of the solar system to 4,568 million years ago, and calculate the formation ages of the very first materials to condense and solidify from the hot gases and dust in the earliest stages of the history of the solar system. Some chondritic meteorites contain materials formed before our solar system was even born, and provide information on the birth and death of stars that existed before our sun. Melted meteorites, the iron, stony-iron and achondrite meteorites, provide us with information and clues on how the first stages of planet formation occurred, and also how bodies such as our moon and the planet Mars formed and evolved over their history. Lunar meteorites allow scientists to study the diversity of the moon's geology; we have meteorites from areas of the moon that were not visited by the Apollo astronauts or the unmanned Luna missions. Although many successful robotic missions have been sent to Mars and returned data on the geology of the planet, no missions have returned actual samples from Mars. Martian meteorites are the only samples of this fascinating world we have to study in laboratories on Earth; they have provided information that is complementary to current space missions, and have played an important part in the planning and development of future missions.

All of these factors and numerous others have allowed scientists to gather evidence to better understand our solar system through its birth and development over nearly 5 billion years of history. If we have a better understanding of our own solar system, we can apply this knowledge to other star and planetary systems within our own galaxy and perhaps beyond. So, in the next few years of meteorite research, what questions remain unanswered or need clarification? What further

OPPOSITE Artist's impression of the European Space Agency's and Russian Roscosmos State Corporation ExoMars rover, which is currently scheduled to land on Mars in 2021. It has a number of scientific instruments and a unique drilling mechanism designed to look for evidence of past or current life on the red planet.

ABOVE An artist's impression of the multi-planetary system TRAPPIST-1. At least seven planets orbit around a central star 40 light years away from Earth, and some of the planets may contain liquid water. Advances in our imaging and understanding of exo-planetary systems will allow us to better understand whether our solar system is unique.

secrets are waiting to be revealed through the dedicated work of researchers around the world? Increases in analytical capabilities should open up a new world to us. Scientists will be able to investigate ever-smaller samples, such as tiny interstellar grains, perhaps composed of just a few tens of atoms of carbon, nitrogen and silicon. The study of this type of material will allow better understanding of the different astrophysical processes that may have occurred, even before the birth of our own solar system. More precise instrumentation will result in the refinement of the chronology of early solar system processes, allowing us to develop more robust timescales to date processes such as accretion and parent-body processes such as aqueous alteration, metamorphism and differentiation.

With many more meteorites waiting to be found in the hot and cold deserts, it is more than likely that we will uncover distinct new meteorite types, broadening the diversity of extraterrestrial material currently known. The current classification of meteorites suggests that there are 'gaps' within and between different meteorite families and groups, so it is possible that the 'missing' link meteorites may one day be discovered.

In parallel with astronomical observations of other planetary systems, we should be able to address the question of whether our solar system is unusual. In particular, why has the solar system been successful in creating a planet that supports life, and is it likely we will find similar life elsewhere?

NASA's OSIRIS-REx mission and Japan's Hayabusa 2 mission are travelling to asteroids that appear similar in composition to the carbonaceous chondrites (see p.56). These missions will combine detailed remote-sensing observations of the surfaces of the asteroids Bennu (OSIRIS-REx) and Ryugu (Hayabusa 2) with collecting samples from the surface for return to Earth and subsequent study in laboratories around the world. Samples from Ryugu are planned to return in 2020 and Bennu in 2023, and scientists are already planning and testing the types of analyses that will be carried out on these precious materials. The composition of organic compounds in these returned samples is of particular interest and is one of the major science questions that these missions will help to answer. As we have seen in Chapter 4, the delivery of organic compounds to the early Earth could have provided the chemical building blocks for the emergence of life. These missions

ABOVE Planned human exploration of the moon, and potentially other solar system bodies, involving long-duration missions will require astronauts to use local resources for survival. This could include being able to manufacture materials for building from local rocks and soils and extracting water from these same materials.

have been carefully designed such that the samples will be protected from potential terrestrial contamination that could interfere with the different scientific studies that are planned.

Other robotic missions are in the planning stages to visit the moon and Mars, to collect samples and return them to Earth. It is hoped that planned lunar sample return missions will visit the south pole region of the moon: the south pole–Aitken Basin. These missions are of interest as this region of the moon contains rocks with a different composition to those sampled by the US Apollo and Russian Luna missions of the 1960s and 1970s, and will provide further information on the formation and evolution of the moon. There is also tantalizing evidence that water ice may be preserved in the soil in the south pole region, and this water would be a valuable resource for astronauts in a future lunar base. Mars sample return is seen as the most complex type of sample return mission and so it is likely to be an international effort, combining the knowledge and technologies of many different space agencies and countries. As with other sample return missions, the samples from a Mars mission will allow for detailed studies in laboratories across the world on materials that have not become contaminated by the Earth's environment. This is particularly important for the key questions related to 'life' on Mars: Is there evidence of fossil life on Mars, or is life existing there today? Mars sample return isa critical step in our continuing study of Mars and for our hopes of sending astronauts to Mars in the 2030s

All these sample return missions will provide a 'ground truth', and allow direct comparisons between remote-sensing observations of these bodies and the wealth of information we have gathered from the study of meteorites derived from these bodies. Both NASA and ESA highlight 'sample return missions' as being important strategic missions as part of our further exploration of the solar system. This feeds back into allowing us to reinterpret what we thought we knew about meteorites in a new light, and will pose 'next-generation' questions.

If we have learned one thing from our study of meteorites it is that for every question or problem that is answered, there are still many, many more to work on. Meteoritics is an evolving science. In many ways, it is still young and in other ways more mature. We've learned a great deal from the study of meteorites over the past 200 years and that knowledge has now driven the evolution of meteoritics from a primarily lab-based, geology–chemistry discipline into one that is part of an exciting future of planetary science, with the ability to visit many of these objects that we've only been able to study indirectly for so many years.

# Further information

**BOOKS**

*A Color Atlas of Meteorites in Thin Section*, D. S. Lauretta and M. Killgore. Golden Retriever Publications, Tucson and Southwest Meteorite Press, 2005.

*Atlas of Meteorites*, M. Grady, G. Pratesi and V. M. Cecchi. Cambridge University Press, 2018.

*Catching Stardust: Comets, Asteroids and the Birth of the Solar System*, N. Starkey. Bloomsbury, 2018.

*Chondrules*, S. Russell, H. Connolly Jr. and A. Krot. Cambridge University Press, 2018.

*David Levy's Guide to Observing and Discovering Comets*, D. Levy. Cambridge University Press, 2003.

*David Levy's Guide to Observing Meteors*, D. Levy. Cambridge University Press, 2007.

*Field Guide to Meteors and Meteorites*, O. R. Norton. Springer, 2008.

*Incoming*, T. Nield. Granta Books, 2012.

*Mars 3-D: A Rover's-eye View of the Red Planet*, J. Bell. Sterling, 2008.

*Meteorites: A Journey Through Space and Time*, A. Bevan and J. R. de Laeter. Smithsonian, 2002.

*Meteorites and Their Parent Planets*, H. Y. McSween Jr. Cambridge University Press, 2nd edn., 1999.

*Meteorites: A Petrologic, Chemical and Isotopic Synthesis*, R. Hutchison. Cambridge University Press, 2007.

*Meteorites: Their Impact on Science and History*, B. Zanda and M. Rotaru. Cambridge University Press, 2001.

*Rocks from Space: Meteorites and Meteorite Hunters*, O. R. Norton. Mountain Press Publishing Company, 1998.

*Roving Mars: Spirit, Opportunity, and the Exploration of the Red Planet*, S. Squyres. Hyperion, 2006.

*Search for Life*, M. Grady. Natural History Museum, London, 2001.

*The Moon*, M. Carlowicz. Harry N. Abrams, 2007.

**WEBSITES**
Note that website addresses are subject to change.

INFORMATION ON METEORITES:

American Museum of Natural History – https://www.amnh.org/exhibitions/permanent-exhibitions/earth-and-planetary-sciences-halls/arthur-ross-hall-of-meteorites

Arizona State University, USA – http://meteorites.asu.edu/

Australian Museum Online – https://australianmuseum.net.au/meteors-and-meteorites

Museum of Victoria, Australia – https://australianmuseum.net.au/meteors-and-meteorites

NASA – http://ares.jsc.nasa.gov/

National Natural History Museum, Paris (in French) – https://www.mnhn.fr/en/collections/collection-groups/mineralogy-and-geology/meteorites

Natural History Museum, London – http://www.nhm.ac.uk/our-science/departments-and-staff/earth-sciences/mineral-and-planetary-sciences.html

### COMETS, METEORS AND FIREBALLS:

American Meteor Society (based in the United States but receives worldwide reports of fireball observations) – http://www.amsmeteors.org/

Fireballs in the Sky (based in Australia but receives worldwide reports of fireball observations) – http://fireballsinthesky.com.au

Fireball Recovery and InterPlanetary Observation Network (FRIPON) (based in France but has information on European observations) – https://www.fripon.org/

International Meteor Agency – http://www.imo.net/

Society for Popular Astronomy, UK – https://www.popastro.com/main_spa1/comet/

https://www.popastro.com/main_spa1/meteor/

UK Meteor Observation Network – https://ukmeteornetwork.co.uk

### THE SOLAR SYSTEM AND SPACE MISSIONS:

**European Space Agency (ESA)**

Aurora – http://www.esa.int/Our_Activities/Human_Spaceflight/Exploration/The_European_Space_Exploration_Programme_Aurora

Luna collaboration with Roscosmos (Moon) – https://www.esa.int/Our_Activities/Human_Spaceflight/Exploration/Luna

Mars – http://exploration.esa.int/mars/

Rosetta (comet) – http://www.esa.int/Our_Activities/Space_Science/Rosetta

**National Aeronautics and Space Administration (NASA)**

Deep Impact (comet) – https://solarsystem.nasa.gov/missions/deep-impact-epoxi/in-depth/

Mars – http://marsprogram.jpl.nasa.gov/; https://www.nasa.gov/topics/journeytomars/index.html

OSIRIS-Rex (asteroid) – https://www.asteroidmission.org

Stardust (comet) – http://stardust.jpl.nasa.gov/home/index.html

**Chinese National Space Administration (CNSA)** http://www.cnsa.gov.cn/

**Indian Space Research Organisation (ISRO)** https://www.isro.gov.in/spacecraft/space-science-exploration

**Japanese Aerospace Exploration Agency (JAXA)**
Hayabusa 2 (asteroid) – http://global.jaxa.jp/projects/sat/hayabusa2/

# Index

# Credits

## PICTURE CREDITS

p.4 ©Pichgin Dimitry/Shutterstock.com; p.7 top ©Matt Genge; bottom ©NASA/JSC; p.8 top©British Antarctic Survey/Science Photo Library; bottom ©NASA; p.12 ©Philippe Thomas; p.13 top left Jon Bodsworth; top right ©Egyptian Museum; p.18 ©totajla/Shutterstock; p.22 © Desert Fireball Network; p.23 top © Desert Fireball Network; p.23 bottom ©UKMON, P.Gilbert; p.24 From American Meteor Society; p.25 top, middle ©NASA; bottom ©Tom Masterson; p.27 ©Photo courtesy of the Antarctic Search for Meteorites Program(ANSMET)/Katherine Joy; p.28 ©Imagery courtesy of the US Geological Survey EROS; p.29 bottom ©Tony Meunier; p.32 ©David Parker/Science Photo Library; p.34 top left NASA; middle, right ©ESA; p.36 ©NASA/NSSDC/Goddard Space Flight Center; p.38 top Svend Buhl / Meteorite Recon [CC BY-SA 3.0 (https://creativecommons.org/licenses/by-sa/3.0)], from Wikimedia Commons; bottom ©Sputnik/Science Photo Library; p.39 ©Ria Novosti/Science Photo Library; p.40 ©Courtesy of Denis Sarrazin, NASA Earth Observatory; p.41 top ©Dr Rob Hough, Western Australia Museum; bottom ©Luigi Folco; p.42 ©Kerry Klein; p.43 issued by the Soviet Union, 20 November 1957; p.44 ©Hans Betlem, Dutch Meteor Society; p.45 ©University of Alabama Museums, Tuscaloosa, Alabama; p.46 ©Walt Radomsky; p.52 top ©NASA, Jet Propulsion Laboratory and ESA, image by C.Carreau; bottom ©JAXA; p.58 Mike Eaton; p.59 ©NASA-JPL; p.60 ©ALMA (ESO/NAOJ/NRAO); p.64 ©NASA Goddard Space Flight Center (NASA-GFSC); p.68 bottom©Martin Lee; p.69©Rhonda Stroud, US Naval Research Laboratory; p.70 ©Keiko Nakamura. Originally reproduced in Zolensky, M E, Tomita, S, Nakashima, S, Tomeoka, K. Hollow organic globules in the Tagish Lake meteorite as possible products of primitive organic reactions. Intl. Journal of Astrobiology, 2002, 1, 179–189; pp.59-60; p.73 ©The Natural History Museum. Redrawn from the original by Cascadia Meteorite Lab and Frank D. Granshaw, Artemis Science; p.80 ©NASA/JPL-Caltech/UCLA/MPS/DLR/IDA/PSI; p.87 ©Luc Viatour; p.89 ©NASA; p.90 Joe Tucciarone; p.91 ©Courtesy USGS Astrogeology Science Center; p.96 ©NASA/JPL-Caltech/LANL/CNES/IRAP/LPGNantes/CNRS/IAS/MSSS; p.99 ©NASA/JPL/Caltech; p.102 ©Jerry Lodriguss/Science Photo Library; p.103 From 'The Midnight Sky', Edward Dunkin; p.104 NASA/JSC; p.105 created using Stellarium.org; p.107 ©Colin Legg; p.108 ©NASA; p.109 top © Photo Scala, Florence; bottom ©Mary Evans Picture Library; p.110 ©NASA/NSSDCA; p.111 ©By NASA and C. Veillet (CFHT) [Public domain], via Wikimedia Commons; p.114 top ©ESA; bottom ©NASA-JPL; p.116 top left ©NASA-KSC; top right ©NASA-JPL; bottom ©NASA-JPL-Caltech; p.117 top left © ESA/Rosetta/NAVCAM, CC BY-SA IGO 3.0; top right and bottom ©NASA/JPL-Caltech/UMD; p.118 © ESA/ATG medialab; p.120 ©ESO/N. Bartmann/spaceengine.org; p.121 ©ESA/Foster + Partners

Unless otherwise stated images copyright of Natural History Museum, London.

Every effort has been made to contact and accurately credit all copyright holders. If we have been unsuccessful, we apologise and welcome correction for future editions and reprints.

## ACKNOWLEDGEMENTS

With thanks to Gretchen Benedix for content retained in this edition, to Tim McCoy (National Museum of Natural History, Smithsonian, USA) for providing very useful comments and suggestions on early versions of the text and also to Malcolm Penn for assistance with GIS and maps.